Coring Operations

The EXLOG Series of Petroleum Geology and Engineering Handbooks

Coring Operations
Procedures for Sampling and Analysis of Bottomhole and Sidewall Cores

**Written and compiled by
EXLOG staff**

Edited by Alun Whittaker

D. Reidel Publishing Company
A Member of the Kluwer Academic Publishers Group
Dordrecht/Boston/Lancaster

International Human Resources Development Corporation • Boston

ACKNOWLEDGMENTS FOR FIGURES

The following figures are reprinted by permission of the Petroleum Extension Service, The University of Texas at Austin, in cooperation with the International Association of Drilling Contractors:

Figures 1–2, 1–3, 1–6, 1–9 are from *Lessons in Well Servicing and Workover, Lesson 2: Petroleum Geology and Reservoirs.* Copyright © 1975 by the University of Texas at Austin.

Softcover reprint of the hardcover 1st edition 1985

Library of Congress Cataloging in Publication Data

Main entry under title:

Coring operations.

(The EXLOG series of petroleum, geology, and engineering handbooks)
Includes index.

1. Core drilling.   2. Drill core analysis.   3. Oil well logging.

I. Whittaker, Alun.   II. EXLOG (Firm)   III. Series.
TN871.2.C647   1985   622'.1828   85–2289
ISBN-13: 978-94-010-8863-3   e-ISBN-13: 978-94-009-5357-4
DOI: 10.1007/978-94-009-5357-4

[90–277–1980–2 D. Reidel]

Published by D. Reidel Publishing Company
P.O. Box 17, 3300 AA Dordrecht, Holland in copublication with IHRDC

Sold and distributed in North America by IHRDC

In all other countries, sold and distributed by Kluwer Academic Publishers Group,
P.O. Box 322, 3300 AH Dordrecht, Holland

*EXLOG is a registered service mark of Exploration Logging Inc., a Baker Drilling Equipment Company.

CONTENTS

ILLUSTRATIONS

PREFACE

This coring operations reference handbook is intended as a practical guide for the logging geologist to procedures, activities, and responsibilities required when bottomhole or sidewall coring is performed at the wellsite. Not all of the operations described are common practice in all logging units; however, familiarity with them is a necessary part of general exploration knowledge and professionalism.

Chapter 1 discusses the concepts of porosity, permeability, and saturation, how these properties are determined in core analysis, and their significance in controlling reservoir performance. Chapter 2 deals with the various techniques used in coring. Chapter 3 explains the routine role of the logging geologist in core retrieval, sampling, and qualitative evaluation. Chapter 4 details operating procedures for quantitative wellsite core analysis equipment.

1

INTRODUCTION

1.1   GENERAL

1.2   QUANTITATIVE CORE ANALYSIS

The primary purpose of coring is to obtain rock samples of a sufficient size to obtain estimates of critical reservoir properties. Core analysis provides direct measurements of:

- Porosity -- the volume of void space contained in a unit volume of rock

- Permeability -- the degree and quality of connection of void spaces which allow fluid flow through the rock

- Saturation -- the relative composition of fluid phases filling the void spaces in the rock (the three principle phases are oil, gas and water)

Unlike wireline log and drillstem test analyses which may also give estimates of these properties, core analysis involves making direct measurements upon a known discrete rock sample. Results are therefore more reliable because there are fewer interpretive assumptions.

Conversely, the process of obtaining the sample (coring, core retrieval, shipping to the laboratory, and sample preparation) causes changes to the physical properties to be measured. Changes in temperature and pressure, and thermal and physical shock may substantially affect porosity, permeability and saturation. Additionally, where substantial secondary porosity exists in the reservoir, whether on a large scale or through uneven distribution, the size of the core sample may be insufficient to represent it. The reservoir properties determined for the core sample, though accurate, may not be directly relatable to whole formation in situ.

Core analysis does not replace the indirect wireline log and DST analyses, but combines with them in developing a complete picture of the reservoir and its productivity.

1.3   QUALITATIVE CORE EVALUATION

Normally of second priority in a decision to cut a core, but of prime importance to the geologist, is the value of a core in providing subsurface stratigraphic information. By their size, known vertical orientation and freedom from contamination and mixing, cores provide reliable observations of

- The physical character of formation boundaries
- Large-scale sedimentary structures of depositional and environmental significance
- Undisturbed and unworked paleontological and palynogical samples
- Uncontaminated geochemical samples for mineralogy and organic geochemistry

Sometimes the requirement for geological evidence may be the overriding factor in the decision to cut a core. For example, on a rank wildcat well with little or no correlative data, it is often necessary to cut a bottomhole core for geological confirmation before making a decision to abandon the well.

1.4 THE ROLE OF THE LOGGING GEOLOGIST

The logging geologist can make a major contribution to the success of a coring operation. By his efforts in performing and supervising core retrieval, sampling, packing and shipping he may contribute to the eventual quantity and quality of data obtained. Paragraphs 1.5 through 1.9 briefly describe the activities of the logging geologist in coring operations which will be covered in the later sections of this manual.

1.5 Core Retrieval

The logging geologist is normally in charge of retrieving a conventional core from the core barrel. Like any other rig floor operation, this must be achieved with the utmost speed and efficiency. You can ensure this by careful and timely preparation.

On the other hand, core retrieval must not be excessively hurried. The object is to retrieve the whole core with minimum breakage and with pieces in the correct order and orientation. You should also make note of the occurrence and location on the core of transient phenomena, like bleeding oil or gas bubbling which may have ceased to occur by the time the core is moved to the logging unit for detailed inspection.

1.6 Core Sampling

Parts of the core will need to be removed and sealed for later quantitative core analysis. This should be done as quickly as possibly after retrieval to preserve contained fluids.

When removing core analysis samples, remember the requirement to preserve the integrity of the core. Avoid breaking the core whenever possible and note the location, length and orientation of pieces removed.

1.7 Core Description

A full description of all geological and hydrocarbon aspects of the core should be prepared for attachment at the bottom of the Formation Evaluation Log. You will be seeing the core in its freshest state. Details visible at this time may be lost or obscured due to physical damage or drying-out during shipping.

1.8 Core Packing and Shipping

The packing and labeling of the core itself, core analysis samples and the core boxes or containers will contribute to the value of the material when received at the laboratory. Maintaining core integrity and minimizing damage are again the prime requisites.

1.9 Core Analysis

Normally, cores will be shipped from the drilling rig, by the fastest means, to a laboratory for core analysis to be performed. In remote locations, long-time delay and the risk of core damage or loss makes this impractical. In order for reliable results to be available in time to make drilling or completion decisions, core analysis must be performed at the wellsite.

Logging companies can provide portable core analysis kits containing equipment and supplies to perform core analysis in the logging unit. This manual discusses the procedures for operation of this equipment, the calculations and log formats required in presenting the results of core analysis.

Core analysis equipment requires a degree of skill in operation somewhat greater than that required for most logging unit equipment. To aid in developing this skill, the core analysis section of this manual is formatted in a step-by-step "cookbook" fashion with numerous illustrations. This is not intended to insult the intelligence of the graduate logging geologist, but to simplify the various stages of an otherwise complex procedure.

1.10   RESERVOIR CHARACTERISTICS

The objective of core analysis is to help reconstruct the most accurate picture possible of the reservoir and its productive potential. By subjecting representative samples from a conventional core or from sidewall cores to the tests described in this manual, it is possible to measure some of the physical properties of the reservoir rock and its contained fluids.

Routine core analysis includes the determination of porosity, air permeability and fluid saturation. Core water salinity and resistivity as well as oil gravity and fluorescence may also be determined. The three major physical properties measured in core analysis tests are

- porosity
- permeability
- saturation

1.11 POROSITY

The porosity of a rock can be defined as the percentage of the total bulk volume of the rock occupied by pore spaces, that is, the portion of the bulk volume not occupied by solid material.

The absolute porosity includes all of the pore spaces, channels and openings of any size, shape or degree of continuity. In the vast majority of sedimentary rocks, a portion of the pore space is isolated -- it is not interconnected with other pore spaces and does not contribute to the productive capacity of the reservoir. The effective porosity includes only the pore spaces that are interconnected and through which fluid can flow.

1.12 PERMEABILITY

The ease with which a particular fluid can flow through the interconnected pore spaces of a rock denotes the degree of permeability to the fluid.

The unit of measurement of permeability, the darcy, was devised in 1856 by Henry d'Arcy, a French engineer. A rock has a permeability of one darcy (1d) when 1 cubic centimeter of fluid of 1-centipoise viscosity flows between the opposite faces of a 1-centimeter cube of the rock in 1 second under a pressure differential of 1 atmosphere. (Water at 68° F has a viscosity of 1-centipoise.)

Because most reservoir rocks have average permeabilities considerably less than 1 darcy, the usual measurement is millidarcies (md, thousandths of a darcy). Permeability of a highly porous, well-sorted sand varies from 475 md for a coarse-grained sand to about 50 md for a very fine-grained sand. Permeability may decrease to about 10 md for a poorly-sorted sand.

Figure 1-1 illustrates the relationship between absolute porosity, effective porosity and permeability.

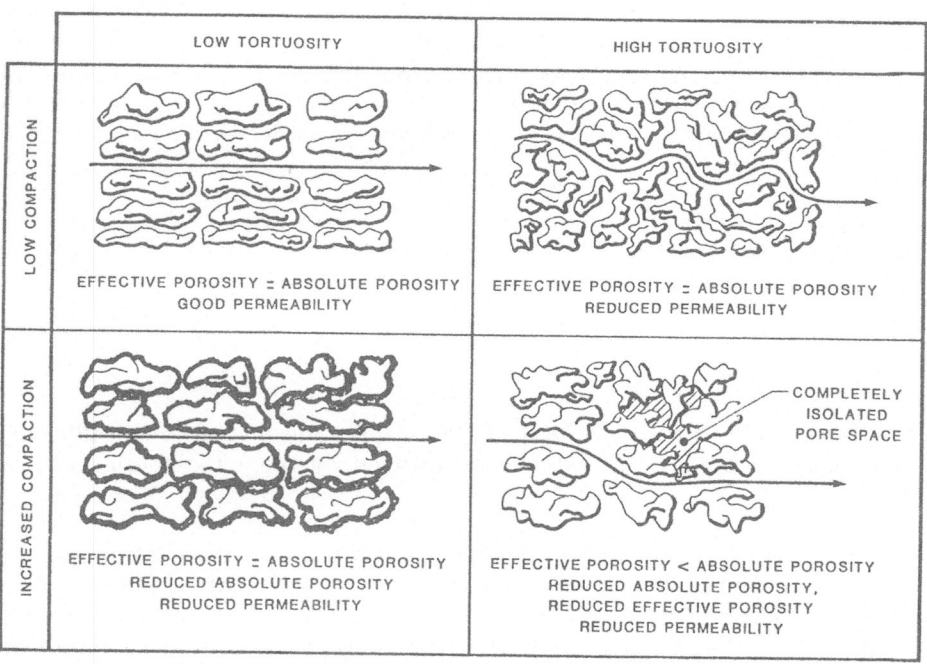

Figure 1-1. Absolute Porosity, Effective Porosity and Permeability

1.13 SATURATION

The term 'saturation' refers to the percentage of the rock pore space that is filled with each of the three fluids -- oil, water and gas. The total of the three will always be one hundred percent, as there is no such thing as empty porosity.

As the core will have undergone changes (such as flushing) while being drilled and will have been subjected to temperature and pressure changes en route to the surface, the saturations measured in the laboratory will probably be quite different to the saturations present in the reservoir itself. In order to fully appreciate the test results of core analysis, a brief description of petroleum reservoirs is included here.

1.14 RESERVOIR GEOLOGY

A petroleum reservoir is a rock capable of containing gas, oil or water, or any combination thereof. To be commercially productive, it must have sufficient thickness, areal extent and pore space to contain an appreciable volume of hydrocarbons, and must yield its contained fluids at a satisfactory rate when penetrated by a wellbore. The most common reservoir rocks are sandstones and carbonates.

The porosity characteristic of a rock may be primary, such as the intergranular porosity of sandstone, or it may be secondary, due to chemical or physical changes such as dolomitization, solution channels or fracturing. Porosity may be reduced by compaction and cementation. The pore space distribution of a reservoir rock is the result of numerous natural processes.

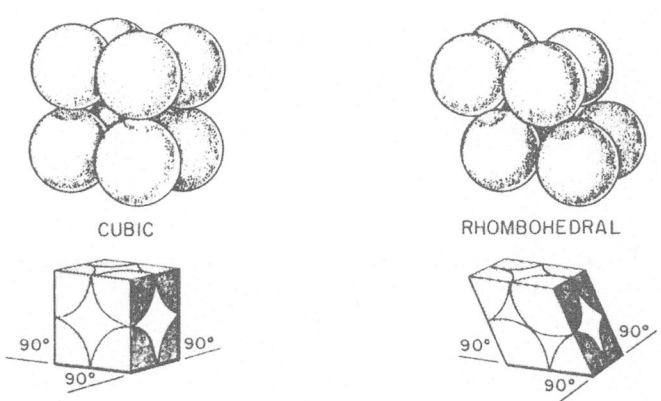

Figure 1-2. Cubic and Rhombohedral Packing

In sandstones, porosity is controlled primarily by sorting, cementing, clay content, and packing. Porosity will be at a maximum when grains are spherical and all of one size, but becomes progressively less as the grains become more angular and less well sorted, because angular grains pack together more closely. Figure 1-2 shows two ways in which

6

spherical grains can be packed. The one on the left is open cubic packing and would have a porosity of about 48 percent. The close rhombohedral packing on the right has a porosity of about 26 percent because the grains are packed into a smaller space. Artificially-mixed clean sand has measured porosities of about 43 percent when well sorted, almost irrespective of grain size, decreasing to about 25 percent for poorly sorted medium- to coarse-grained sands. Very fine-grained sands still have more than 30 percent porosity. Figure 1-3 summarizes diagrammatically some of the processes controlling pore size distribution in sandstones, and Figures 1-4 and 1-5 detail diagrammatically the reduction in porosity due to compaction and cementation.

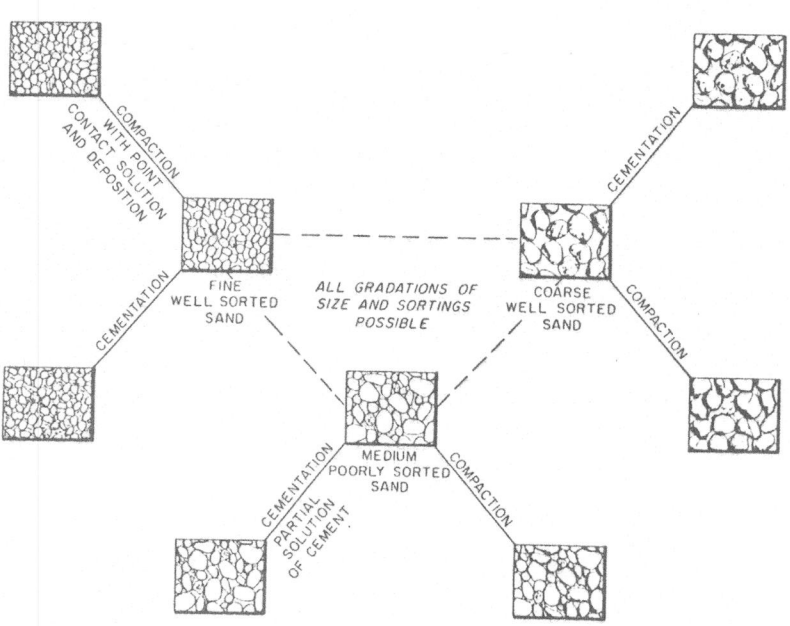

Figure 1-3. Pore Size Modification in Sandstones

Compaction caused by weight of the overburden packs the sand grains closer together, and at greater depths may even crush or fracture the grains. The result is smaller pores and therefore lower porosity -- but more importantly, a much lower permeability. Thus a sandstone reservoir which is capable of producing petroleum at 10,000 feet may be too impermeable at 20,000 feet to be economicallly productive. Cementation, which will fill part or all of the pore space, also tends to increase with depth. Figure 1-6 shows the effect of compaction on a poorly sorted sandstone from Ventura, California. The silhouettes are based on photomicrographs but are easier to see in black and white, as shown. Note how the pore space diminishes from surface conditions through the zones to the Miocene. There is still quite a bit of porosity in the Miocene, but the permeability has decreased greatly.

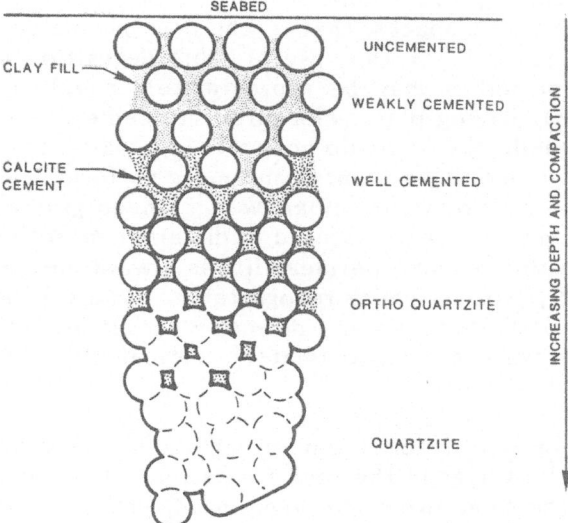

Figure 1-4. Porosity Reduction with Compaction and Cementation

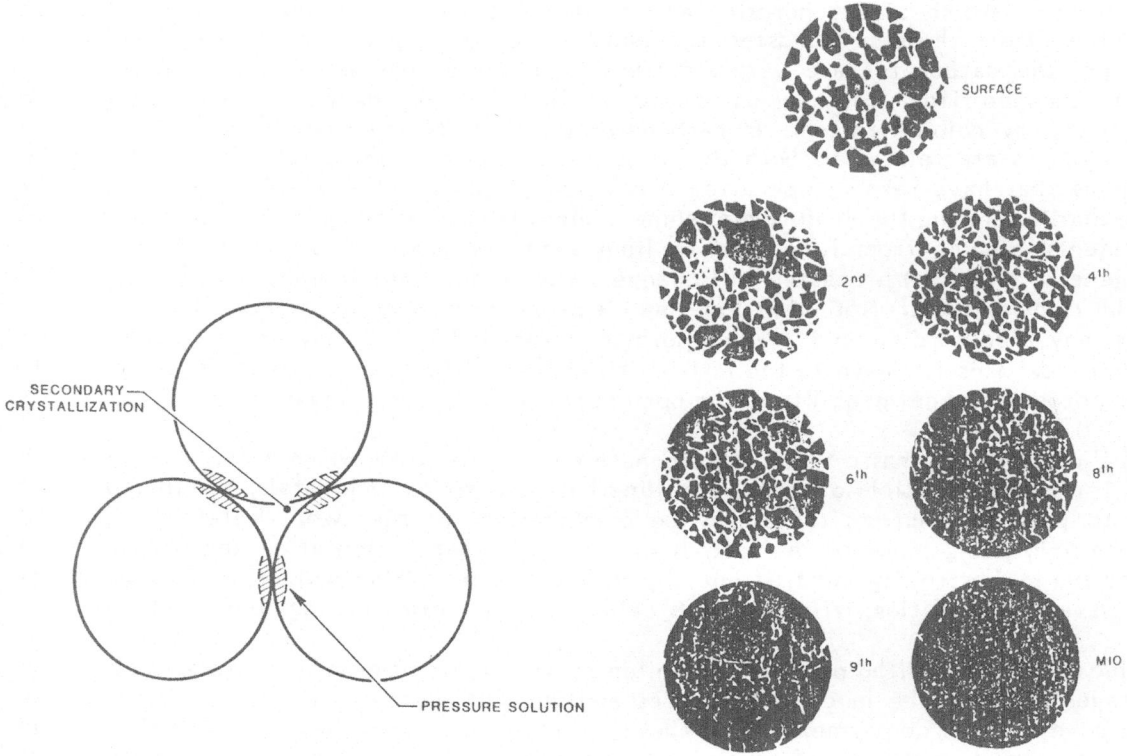

Figure 1-5. Pore Size Reduction with
Solution and Recrystallization

Figure 1-6. Typical Porosity
Decrease with Depth

The orientation of mineral grains within the rock itself is important as it determines the shape and orientation of pore spaces, that is, the pore geometry of the rock. In the case of well-rounded grain sands, or in the case of rapid deposition where no orientation has been established, permeability may be approximately equal in all directions. At the other extreme, minerals having platy, elongated or needle-like grains are deposited in a position of greatest stability; this would normally be parallel to the bedding planes. As a result, permeability is found to be greater and more uniform in directions parallel to the bedding. Except in the case of a very clean, well-rounded-graine sandstone, variations in permeability will be evident when measured in different directions. In the core analysis test described later in this manual, permeability is always measured in a direction parallel to the planes of deposition, where recognizable. Where bedding planes are not discernible, permeability is determined in a direction perpendicular to the axis of the core. Since most exploration wells are approximately vertical, this is assumed to be horizontal permeability.

Permeabilities vary not only in different directions but may vary greatly in the same direction at different localities in the rock formation. It is not uncommon to find varying permeabilities in the same direction, even in adjacent core samples. Erratic depositional processes can result in the occurrence of localized lenses of so-called 'loose' or 'tight' sands within a formation of otherwise uniform permeability. Permeability may also vary laterally within a field -- very permeable sands occurring in one portion of the field, grading to tight sands in another portion.

The factors which affect porosity and permeability in limestones and dolomites are different from those of sandstones. Due to their susceptibility to post-depositional change, the carbonates are very different from sandstones and shales -- particularly when changed from calcium carbonate to the calcium magnesium carbonate form (dolomite) by dolomitization. In carbonates the porosity, permeability, and pore space distributions are related to both the depositional environment of the sediment and the changes that have taken place after deposition (Figures 1-7 and 1-8). Figure 1-9 is a diagram illustrating the evolution of some dolomite textures and pore types. The original sediments ranging from lime mud to lime sand are depicted in the central column. Diagenetic (or later) processes will change porosity and pore size distribution, as shown to the right and left. Note that lime mud is preferentially dolomitized. The other particles may then be dissolved, leaving pores or larger holes that may or may not be interconnected. This is shown to the left of the center in the figure. To the right, note the good porosity and permeability in the open network of dolomite crystals.

This illustrates two basic types of carbonate porosity (also shown in detail in Figures 1-8 and 1-10): interparticle or intercrystalline between grains or crystals, and intraparticle due to particle solution; there may be combinations of the two. Referring again to Figure 1-9, going more to the left, it can be seen that cementation and infilling have taken place, destroying most, if not all, of the porosity. When volume reduction occurs due to recrystallization, irregular voids called vugs are formed (vuggy porosity).

Permeability in all lithologies is controlled by the size of the pore throats, such as the passages between the much larger pores and vugs. Consequently, a highly porous rock may have little or no permeability if these interconnections are very small or absent. On the other hand, some very-fine-grained carbonate rocks have an extensive network of interconnected pore space with enough permeability to yield economic volumes of oil. Intercrystalline pores tend to be interconnected, and rocks with high intercrystalline porosity are normally permeable as found in many highly productive dolomite reservoir rocks.

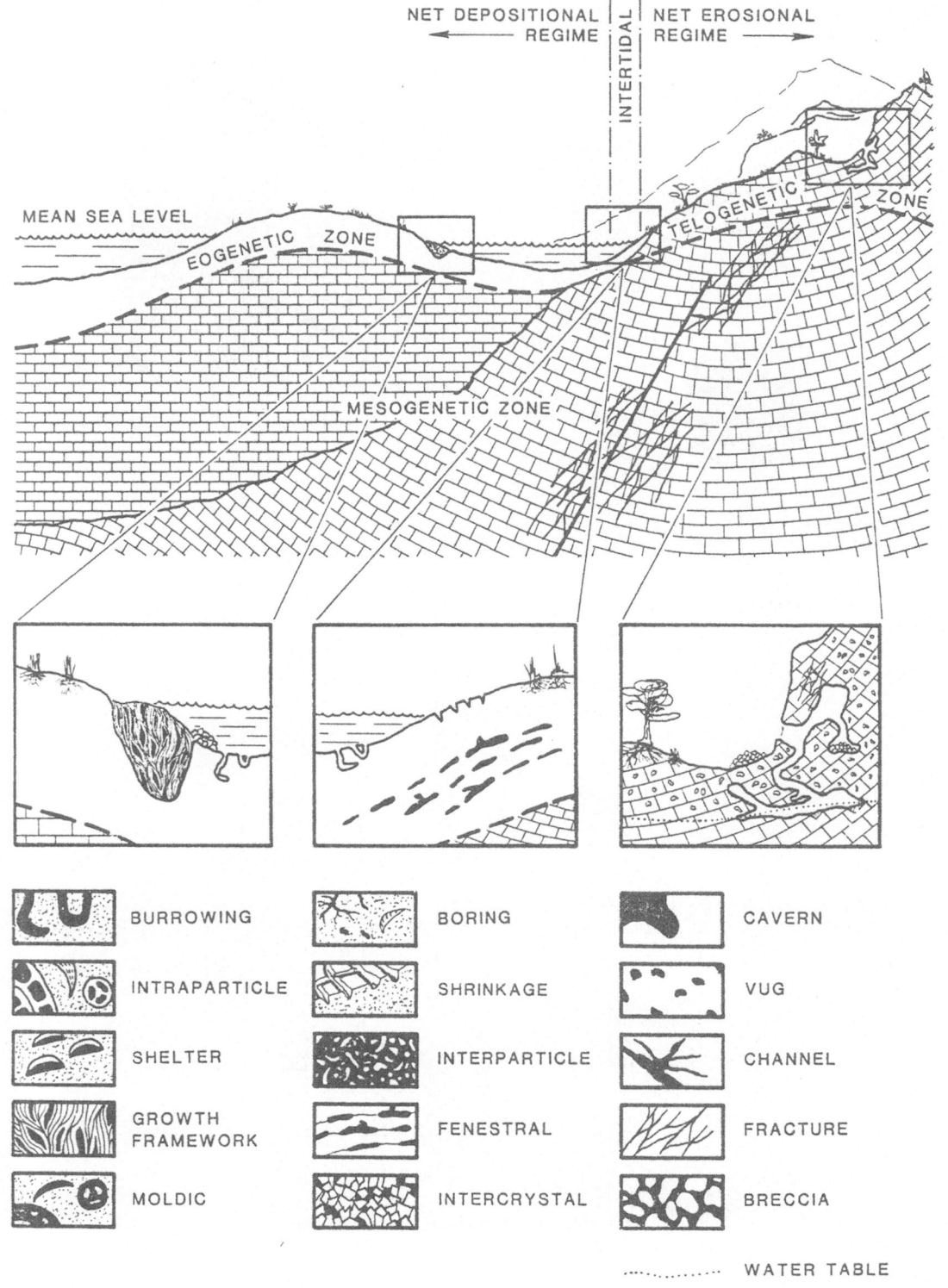

Figure 1-7. Carbonate Porosity Types

10

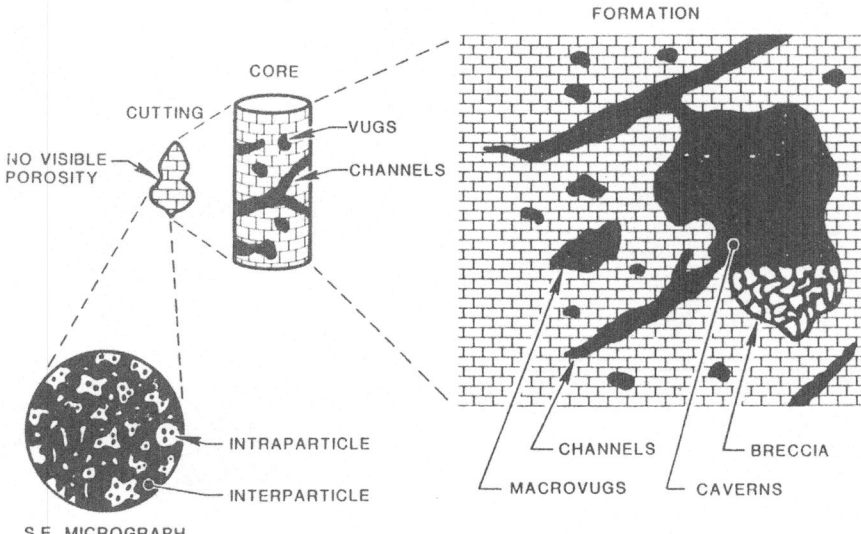

Figure 1-8. Carbonate Porosity Types and Scale

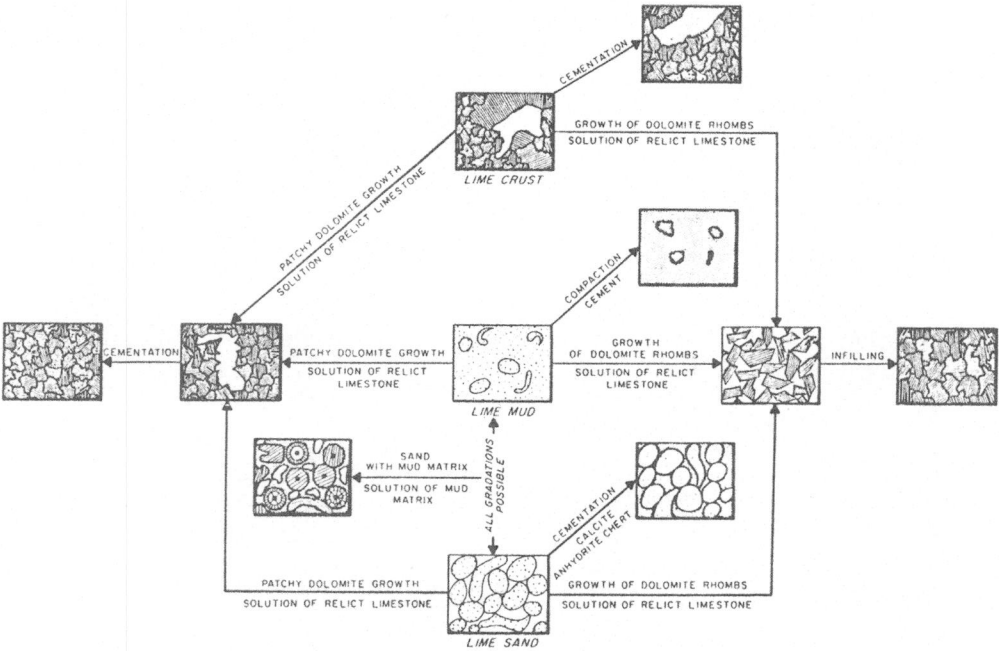

Figure 1-9. Porosity Modification in Carbonates

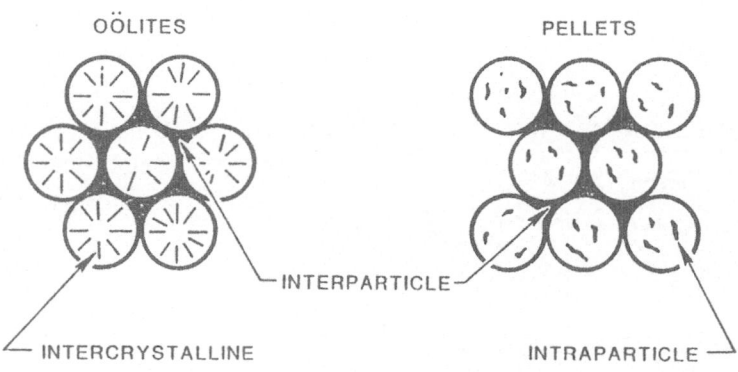

Figure 1-10. Carbonate Interparticle and Intraparticle Porosity Types

Carbonates can be extensively fractured. In this situation, even without porosity and permeability in the main body of the formation, economic amounts of oil can exist in the fractures if the source and other conditions of accumulation are present.

The permeability of a formation is a function of the pore pattern and of the size, shape and distribution of the small pore channels which connect the larger pores, vugs, fractures, etc. The larger and more numerous the pores and channels, the greater the permeability. The size of the connecting channels is important, as small channels restrict flow due to greater frictional losses. Also, the rate of discharge through a cross-section of formation will depend on the nature of the fluid(s) -- in particular, viscosity and the pressure differential. As was seen in Figure 1-1, permeability is not necessarily directly related to effective porosity, although the size, shape and continuity of the pores affect the value of both. In reservoirs of commercial importance, permeabilities can vary from less than 1 to 20,000 millidarcies or more, though the majority ranges between 20 and 2000 millidarcies.

In a potential reservoir where a mixture of fluids is present (water and/or oil and/or gas), the reservoir will have less effective permeability to each fluid than it would to an individual fluid if it were present alone. Some attempt to reconstruct the character of the fluids in place is necessary in estimating the nature of the eventual production from the reservoir.

Relative permeability can be defined as the effective permeability of a reservoir to a fluid when one or more other fluids are present. The effective permeability to each phase will be a function of the number of phases present, the proportion of pore volume occupied by each phase, the chemical composition and physical properties of each phase, and the degree of continuity of each phase (Figure 1-11). It is therefore important to understand how fluids are distributed in a reservoir, and also the phases (liquid or gas) in which each exists under the prevailing conditions of temperature and pressure within the reservoir rock.

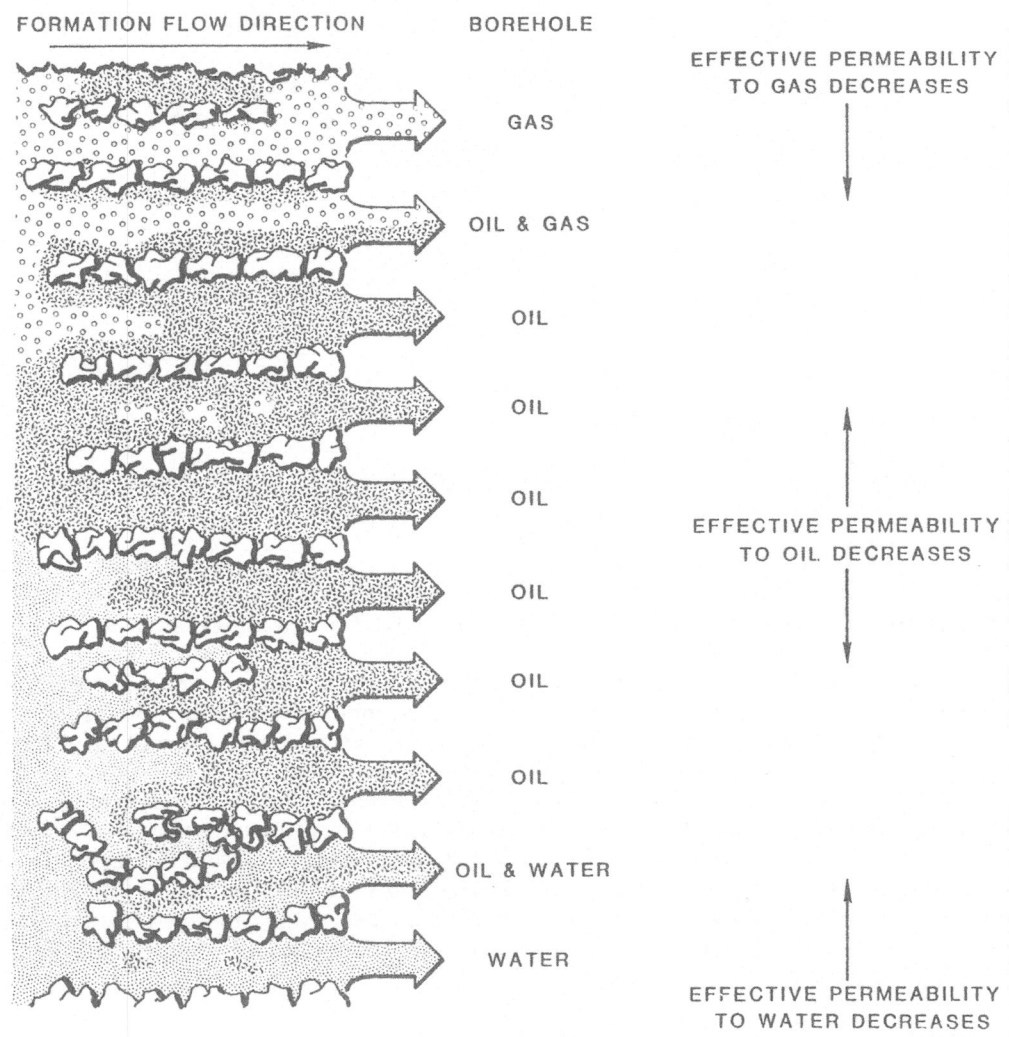

Figure 1-11. Relative Permeability and Productivity

1.15 DISTRIBUTION OF FLUIDS

Figure 1-12 shows an idealized petroleum reservoir. The fluids are distributed by gravity segregation; water occurs in the lower portion of the structure, gas in the top, and oil is intermediate between the two. Although gravity plays an important role in determining the distribution of fluids, other factors have an appreciable influence, especially in the distribution of the liquid and gas phases. In actual reservoirs, water is nearly always present within the oil zone as a thin film on the sand grain surfaces and occupying the finer capillary pores. This water is known as connate water -- originally present in the reservoir before the migration and accumulation of oil. The amount may vary from a small percentage in some reservoirs to more than 50 percent in others. If consideration is given to the interfacial and displacement forces active at the time the oil first migrated into the reservoir, it would be expected that the oil displaced only that water present in the larger pore spaces. The surface tension of the water prevents oil from displacing the surface film of water on the grain surfaces and the water present in the capillary-size pores.

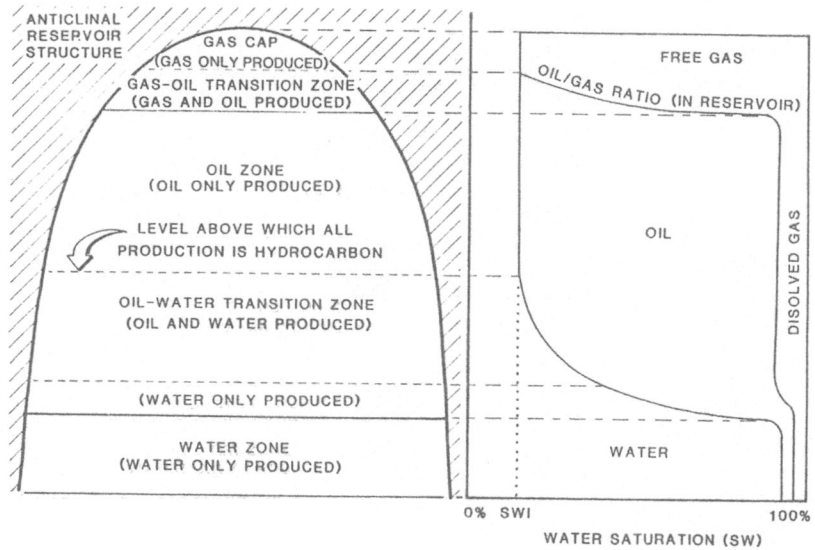

Figure 1-12. Idealized Reservoir Composition

Fluid properties such as (1) the interfacial tension between reservoir oil and connate water and (2) the density difference between these fluids are factors which, together with the rock properties, determine the initial static fluid distribution in a reservoir and hence the thickness of the transition zone.

In the reservoir, water occupies the smallest pores and the corners and crevices of the larger pores, while the hydrocarbons occupy the center of the larger pores. The amount of interstitial or connate water saturation depends on the pore geometry (and indirectly, the permeability), the height above free water level (if present), and the properties of the fluids. In going onward from a water-oil contact in a formation of uniform permeability, the oil saturations increase rapidly, the water saturations decrease correspondingly through a transition zone, and the connate water reaches a near-minimum value (termed the irreducible water saturation). Oil alone will be produced from this zone. In a homogeneous formation of moderate to high permeability, the transition zone is ordinarily quite short, that is, up to several feet in thickness. In uniform but low-permeability formation, the transition zone is larger.

If the permeability is very irregular and there are streaks of abnormally low permeability interspersed in permeable formation, the effective length of the transition zone may be increased. A low-gravity or high-viscosity oil tends to produce the same result as an erratic permeability distribution, since the viscosity condition magnifies the importance of the zones of lower permeability and higher water content. If the reservoir oil is unusually heavy (approaching the density of water), another factor comes into play, as a small density difference in itself tends to create a long transition zone.

Figure 1-13 illustrates the type of fluid that would be produced from each of the zones in our idealized reservoir.

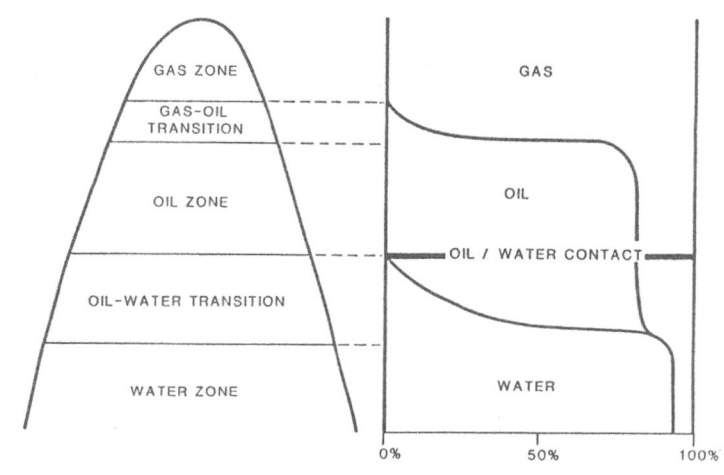

Figure 1-13. Idealized Reservoir Productivity

1.16 PHASE RELATIONSHIPS

Other important factors when considering the fluids in place in a reservoir are the pressure and temperature differences between the reservoir itself and the surface -- and in particular, the phase relationships between the gas and liquid states at varying temperatures and pressures. Gas which is detected from the mudstream and isolated as originating from the formation need not be present in the formation as free gas.

In Figure 1-14, segregation between liquid and vapor is shown for two single pure hydrocarbons under varying conditions of temperature and pressure. For each hydrocarbon there is a fixed relationship between temperature, pressure and phase. These are unchangeable and may be determined both empirically and by derivation from the Gas Laws. Examination of the 'Vapor Pressure Curves' for the pure hydrocarbons shows that, at any fixed temperature below the Critical Temperature (T_c, a constant for any pure substance), application of pressure to the gas results in liquification. Conversely, reduction in pressure results in evaporation. Alternatively, at any fixed pressure below the Critical Pressure (P_c, the pressure of the saturated vapor of a substance at its Critical Temperature), increasing temperature results in evaporation, decreasing the temperature in liquification. Above the Critical Point defined by Tc and Pc, changes in temperature and pressure do not affect the state of the substance which may be considered to be in a single, indeterminate phase. This phase may be referred to as a super-heated liquid or a super-saturated vapor.

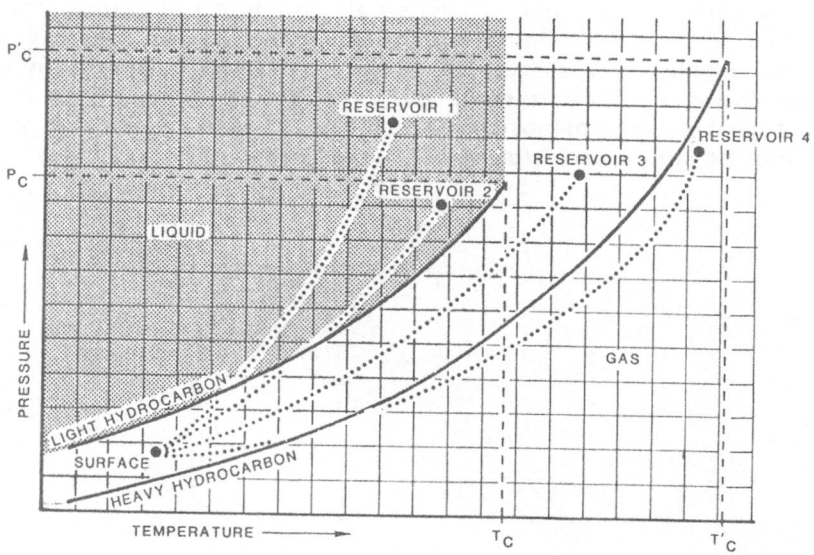

Figure 1-14. Phase Distribution for Single Pure Hydrocarbons

The importance of this to reservoir evaluation is shown in Figure 1-14 by the points which indicate surface conditions and those from four typical reservoirs. As the hydrocarbon travels from the reservoir to surface, it undergoes changes in temperature and pressure, with possible accompanying change in phase. The light hydrocarbon can be present in a liquid phase in Reservoir 1 but may evaporate and be detected as a gas at the surface. Conversely, the heavy hydrocarbon will be present as a gas in Reservoir 4 but will condense to a liquid at the surface. The other examples in the illustration show other possible changes or retentions of state. The importance of these facts in the interpretation of gas shows and of the core analysis test results is more than simply that, a gaseous hydrocarbon in place may be evaluated as a liquid at surface. (It is, after all, at surface that the hydrocarbons will eventually be produced and processed. The reservoir engineer is quite able to extrapolate from surface-to-reservoir conditions and compositions.) The important factor is that the physical properties of the liquid and gaseous phases are markedly different. Thus, factors such as viscosity and the reservoir's permeability to a particular hydrocarbon, -- in sum, its mobility in the reservoir -- are dependent on the phase in which the hydrocarbon is present. Thus, two reservoirs containing similar fluids may show markedly different production rates or types due to variation in reservoir temperature and pressure.

This is a simplification indicating the different behavior of two pure hydrocarbons when liberated from various reservoir conditions and released at surface. A reservoir fluid will never be entirely a pure, single hydrocarbon (although, of course, non-associated gas which may consist of 90 percent or more of methane is close to this). Most reservoir fluids will be complex mixtures of varying composition. In such a mixture the temperature-versus-pressure relationship is also complex (Figure 1-15).

Instead of a single curve indicating the division of the liquid and gaseous states, there are two curves, known as the bubble-point and the dew-point curves, meeting at the critical point. Above the bubble-point curve, the mixture will exist only as a liquid. Below the dew-point curve, only gas will be present. Between the curves, mixtures of liquid and gas will be present together, the composition of each phase being a function of the temperature, pressure and the net composition of the total fluid.

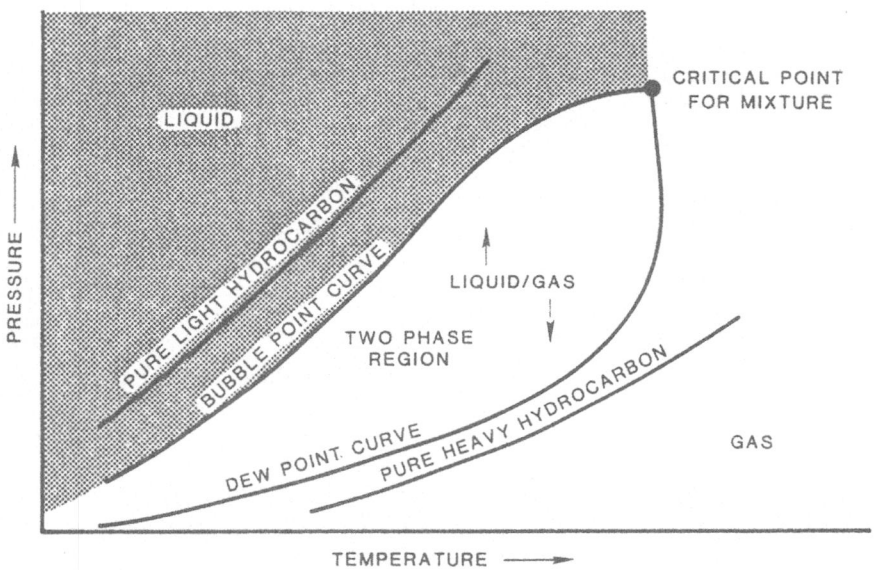

Figure 1-15. Phase Distribution in a Two-Component Hydrocarbon Mixture

It is also worth noting that the bubble-point/dew-point curves of a two-component mixture will be enclosed by (but not coincide with) the vapor pressure curves of the two pure components. Therefore, a two-component mixture, at a given temperature and with falling pressure, will remain totally liquid to a pressure below that at which the pure, lighter component would become a gas. Further decline in pressure would lead to the bubble-point below which the mixture would have a declining liquid phase and an increasing gaseous phase, until the dew-point were reached at which the mixture would be totally gaseous. This would occur at a pressure higher than that at which the pure, heavier component would become entirely gas.

From the above, it can be seen that the quantity of gas and oil present in a reservoir under prevailing conditions is a function of temperature, pressure, the number and amount of hydrocarbon species present, and the physical properties of each species. This is further complicated by the presence of water as both a liquid and vapor phase and by the limited though significant mutual solubilities of water and hydrocarbons.

1.17   CORE CHARACTERISTICS

Core analysis involves the measurement of porosity, permeability and saturations of samples taken along the length of a core. When performed carefully, upon good samples, these analyses can produce a reliable representation of these characteristics through the whole core.

Regardless of the care and reliability of analyses, however, the results obtained can only provide a partial indication of the characteristics of the reservoir in situ. This is in part due to the complexity and heterogeneity of the reservoir as a whole, as described in Paragraphs 1.10 through 1.16. It also results from the predictable but non-quantifiable changes undergone by the core in the process of cutting and retrieval to surface.

In Paragraphs 1.18 through 1.26, we discuss the actual methods used in core analysis and the way in which their results are comparable to in situ reservoir characteristics. Specific operating and analytical procedures are discussed in Section 4.

1.18 POROSITY

1.19 Measurement

The standard instrument for measurement of porosity is the Ruska Porometer (Figure 1-16). It consists of

- A Pycnometer -- a container of precisely known internal volume

- A Mercury Pump -- with which precisely measured volumes of mercury (a relatively incompressible fluid) may be introduced into the pycnometer

- Vernier Scales and Pressure Gauges -- used to determine the pumped volumes

The porometer is used to accurately determine the volume of samples of core, normally one inch diameter cylindrical samples known as plugs. A plug is first washed with solvent and dried to remove all reservoir fluids. It is then placed in the pycnometer. Mercury is pumped into the pycnometer to a constant pressure. The difference in volume of mercury that can be pumped into the pycnometer empty and when it contains the core plug is equal to the volume of the plug.

$$V_c = V_{m_1} - V_{m_2} \tag{1-1}$$

where

V_c =   volume of core plug, cc

V_{m_1} =   volume of mercury pumped into empty pycnometer, cc

V_{m_2} =   volume of mercury pumped into pycnometer with plug, cc

If pumping continues, the volume of air trapped within the connected pore space of the plug will compress. Using Boyle's Law, which relates pressure and volume, it is possible to determine the volume of trapped air from the increase in pressure in the pycnometer as a measured extra volume of mercury is pumped in. Thus

$$V_c = V_g + V_p \tag{1-2}$$

$$\phi_e = \frac{V_{pe}}{V_c} \tag{1-3}$$

where

V_g =   grain volume of pore plug, cc

V_{pe} =   effective pore volume of core plug, cc

ϕ_e =   effective porosity of core plug, fractional (0-1.0)

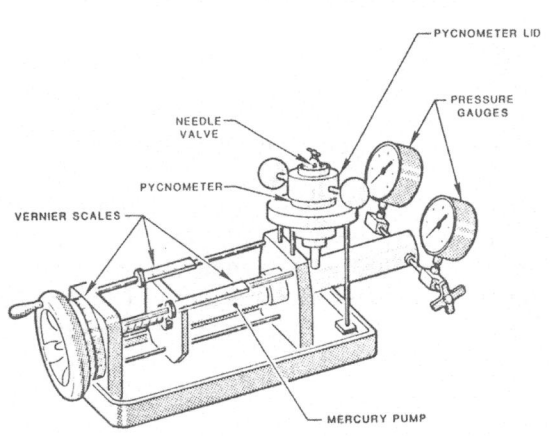

Figure 1-16. Ruska Porometer

Since sealed, unconnected voids in the plug will remain isolated from the external imposed pressure, they will be unaffected and undetected by this measurement. The result, therefore, is Effective Porosity, the connected void space in the rock.

Combining the porometer results with other measurements can give estimates of Absolute Porosity, the total void space in the rock. For example, if the rock is relatively clean and consists of a single mineral of known density, the mass of the plug can be determined by weighing.

$$\rho_b = \frac{Mc}{Vc} \tag{1-4}$$

$$\phi_a = \frac{\left(\rho_m - \rho_b\right)}{\left(\rho_m - \rho_f\right)} \tag{1-5}$$

where

Mc = mass of core plug, gm

ρ_b = bulk density of rock, gm/cc

ρ_m = matrix density of rock mineral, gm/cc

ρ_f = fluid density of pore fluid, gm/cc

ϕ_a = absolute porosity of core plug, fractional (0-1.0)

If the fluid density, air at atmospheric pressure, is assumed to be zero, this simplifies to

$$\phi_a = 1 - \frac{\rho_b}{\rho_m} \tag{1-6}$$

A more accurate estimate, especially when the rock consists of mixed lithologies of varying density, is to determine rock grain volume. This is done by crushing the plug to individual grains and determining the volume of grains using a glass volumeter (Figure 1-17).

$$\phi_a = \frac{V_c - V_g}{V_c} \qquad (1-7)$$

where

V_c = volume of core plug (from porometer), cc

V_g = grain volume of core plug (from volumeter), cc

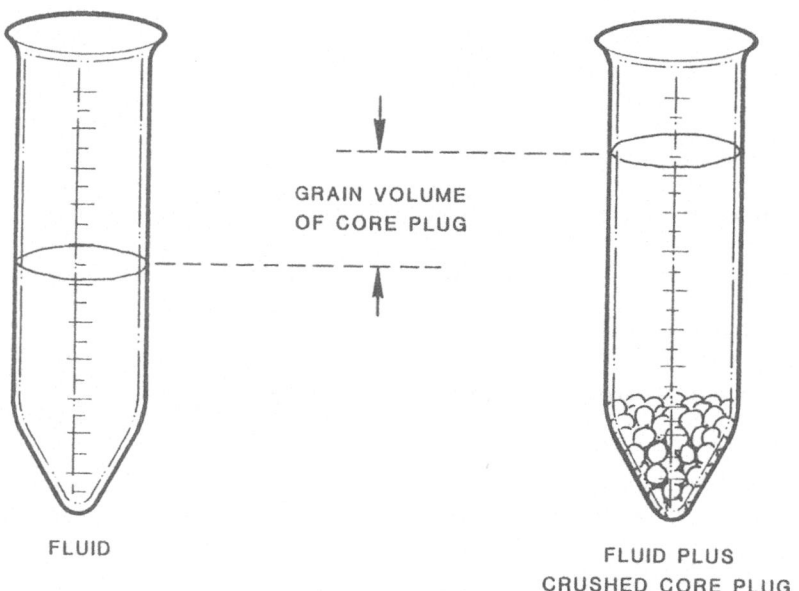

FLUID

GRAIN VOLUME
OF CORE PLUG

FLUID PLUS
CRUSHED CORE PLUG

Figure 1-17. Glass Volumeter for Determination of Absolute Porosity

1.20 Limitations

Porosity determination on a 1-inch diameter plug may only be representative of porosity types, the size and distribution of which are adequately represented in so small a sample (that is, primary, matrix-controlled porosity). In clastic rocks this may be the most important and volumetrically significant porosity. In carbonates, secondary porosity, fractures and diagenetic solution features may be volumetrically predominant. In such cases, the size or distribution of such porosity may be on so large a scale that a core plug or even a whole core cannot represent it. Alternatively, organic porosity may be on so small a scale that even pulverized rock contains some porosity (see Figure 1-8).

Where porosity size and distribution are of a scale that may be represented in a core plug (for example, intergranular porosity in a fine-grained sandstone), reasonable porosity estimates are possible. Even so, the porosity estimate from the core plug will reflect both the true in-situ rock porosity and the results of porosity-modifying mechanisms occuring during the cutting and retrieval of the core. These are of two types:

- Porosity reduction by clay hydration
- Porosity enhancement by microcrack formation

During drilling, a permeable formation ahead of the drill bit is extensively flushed with mud filtrate. In coring, with substantially lower rates of penetration, the duration and hence extent of this flushing are much greater. We see this effect reflected in ditch gas reading during coring operations. Even when accounting for the lower rate of penetration and smaller volume of cut formation, ditch gas readings when coring are lower than expected (see Figure 1-18).

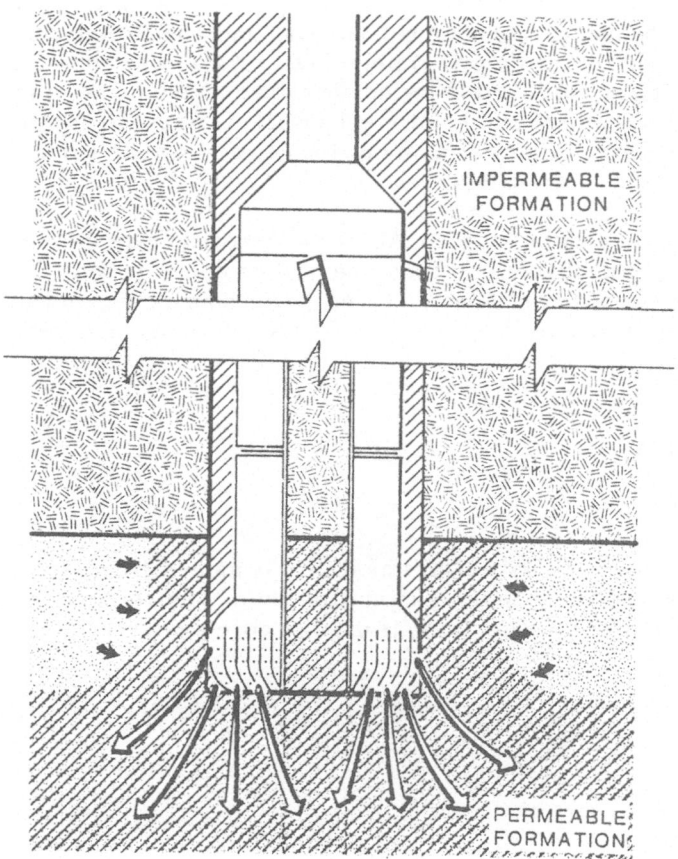

IMPERMEABLE FORMATION

PERMEABLE FORMATION

Figure 1-18. Flushing of a Permeable Formation while Coring

In addition to sweeping fluids from the core, flushing brings mud filtrate into contact with clays dispersed with the formation porosity. Hydration and expansion of these clays may result in an overall, absolute porosity reduction. Where expansion occurs in restrictions and pore throats, the result may be in reduction permeability or, in the extreme, closure of the pore throat, rendering the porosity ineffective. Although later drying of the core sample may reverse some of these effects, it is likely that some permanent change will occur in either absolute porosity, effective porosity or both. The significance of these changes will depend upon the original porosity and the quantity, type and distribution of clay in the rock.

When a core is cut in a well-indurated rock it is common, on recovery, to discover major subvertical joint-like cracks in the core. The orientation of these cracks shows no relationship to that of other rock features or texture. Examination of crack surfaces shows no evidence of movement, fluid flow, secondary deposition or healing. The cracks are, therefore, interpreted to be newly formed as a result of stress relief in retrieval of the core and not a true feature of the formation. Obtaining core plugs for porosity determination can be difficult. Plugs may break or contain cracks leading to falsely high porosities.

Recent evidence from comparison of analyses from conventional and pressurized cores (see Section 2) indicates that the core cracking phenomenon is not limited to well-indurated rocks. Strickland et al (SPE 8303, 1979) reported extensive cracking on a microscopic scale in low porosity sandstones. The microcracks usually consisted of the rupturing of grain boundary cementing material and the production of cracks in shale intercalations in a horizontal (bedding) plane. Cracks across quartz grains were isolated except in the vicinity of major cracking as described above. The cracks were interpreted as resulting from stress relief and from mechanical stresses during the cutting, retrieving and handling of the core. Thermal contraction in retrieval to surface was considered to be a minor contribution to cracking.

The contribution of microcracking to Absolute Porosity may be very small. The total microcrack porosity is probably less than one half of one percent. On the other hand, the microcracks which form at previously sealed grain boundaries and in clay pore fill may contribute to a large relative increase in Effective Porosity and Permeability (see Paragraph 1.23).

1.21 PERMEABILITY

1.22 Measurement

Henry D'Arcy, for whom the unit of permeability is named (see Paragraph 1.12), proposed a series of equations defining various modes of fluid flow through porous media. One of these, assuming horizontal, linear flow, is used in core analysis:

$$Q = \frac{KA \; \Delta P}{\mu \; L} \qquad\qquad (1\text{-}8)$$

$$K = \frac{Q \; \mu \; L}{A \; \Delta P} \qquad\qquad (1\text{-}9)$$

where

Q   =   flowrate, cc/sec

A   =   cross sectional area of flowpath, cm^2

ΔP =   pressure differential across flowpath, atm

L   =   length of flowpath, cm

μ =   fluid viscosity, cps

K   =   permeability, darcies

The standard instrument for measurement of permeability is the Ruska Permeameter (Figure 1-19). It consists of

- A sample holder, in which air flows through a cylindrical core plug
- Thermometer
- Air flowmeters
- Air pressure gauge

A core plug is solvent-washed and dried to remove reservoir fluids. Its length (L) and diameter are measured and the cross-sectional area (A) calculated. The plug is then sealed in the sample holder and air is forced through it at a measured pressure (ΔP) and flowrate (Q). The temperature of the air is measured and used to calculate its viscosity (μ). Substituting these known and calculated values into equation (1-9) allows air permeability of the rock to be calculated.

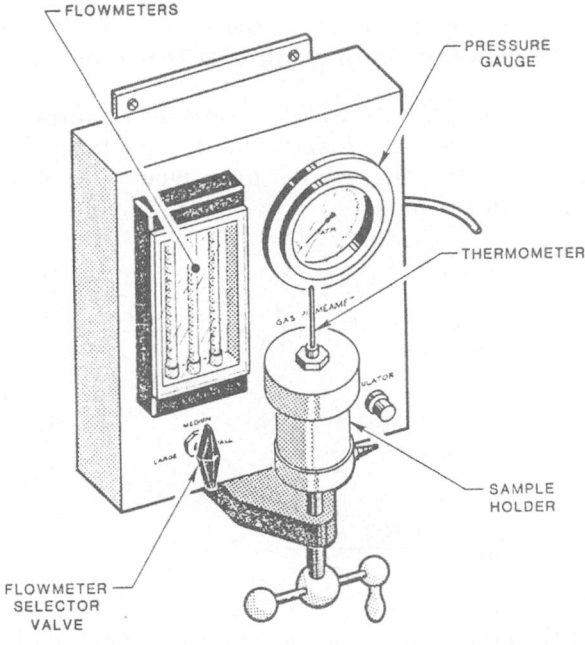

Figure 1-19. Ruska Permeameter

1.23 Limitations

All of the limitations of core analysis porosity estimates (Paragraph 1.20) apply to an even greater extent to permeability estimates. Factors of scale and porosity modifying mechanisms will affect permeability as explained, but to an extent many times that of the magnitude of the porosity change.

Although many factors control the relationship between porosity and permeability, it can be generalized in the form

$$K = Ae^{\phi} - B \tag{1-10}$$

where

K $\quad$ = $\quad$ permeability

ϕ $\quad$ = $\quad$ porosity

A, B $\quad$ = $\quad$ constants

From this equation, it can be seen that only a small change in porosity will produce a much greater change in permeability. For example, the minute increase in porosity introduced by microcracking may result in large increases in measured permeability. Strickland's observations show that most of the induced porosity is associated with microcracks aligned in a horizontal plane. This is the direction in which core plugs for permeability measurements are normally cut (see Paragraph 4.9).

Another factor which influences the reliability with which permeability estimates may be applied in deducing reservoir performance is the way in which permeability variations within a sample affect the measured permeability of the whole sample.

Figure 1-20 illustrates two core plugs in which variations in porosity and permeability are represented by division in two halves, longitudinally and laterally. In either case, the porometer-measured porosity for the whole plug would be a simple average of the volumes of porosity present in the plug:

$$\text{Whole Plug } \phi_m = \frac{\left[\frac{LA}{2} \cdot \phi_1 + \frac{LA}{2} \cdot \phi_2 \right]}{LA} \tag{1-11}$$

$$= \frac{\left(\phi_1 + \phi_2 \right)}{2}$$

$$\phi_m = \frac{20\% + 5\%}{2} = 12.5\%$$

The reliability of this estimate depends upon how well the sample provided by the core and plug represents the whole formation.

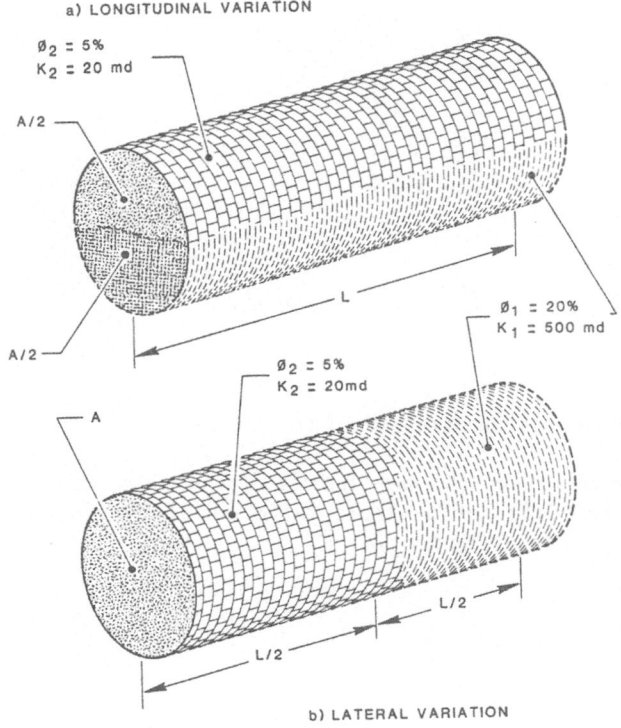

a) LONGITUDINAL VARIATION

$\emptyset_2 = 5\%$
$K_2 = 20$ md

$A/2$

$A/2$

L

$\emptyset_1 = 20\%$
$K_1 = 500$ md

$\emptyset_2 = 5\%$
$K_2 = 20$md

A

$L/2$

$L/2$

b) LATERAL VARIATION

Figure 1-20. Permeability Variation within a Core Sample

Permeameter-measured permeabilities would be different for the two plugs. In case A, the longitudinal variation, the upper and lower halves have different permeabilities and would sustain different air flowrates under the applied differential pressure. Total flowrate through the whole core would be the sum of these two:

$$Q = Q_1 + Q_2$$

$$= \frac{K_1 \frac{A}{2} \Delta P}{\mu L} + \frac{K_2 \frac{A}{2} \Delta P}{\mu L}$$

$$= \frac{\left(K_1 + K_2\right) \frac{A}{2} \Delta P}{\mu L} \qquad (1\text{-}12)$$

Substituting this flowrate into equation (1-9) we may determine the measured permeability.

$$\text{Whole Plug } K_m = \frac{Q \mu L}{A \Delta P} \qquad (1\text{-}9)$$

$$= \frac{(K_1 + K_2) \, \frac{A}{2} \, \Delta P}{\mu L} \cdot \frac{\mu L}{A \, \Delta P} \tag{1-13}$$

$$= \frac{(K_1 + K_2)}{2}$$

$$K_m = \frac{500 + 20}{2} = 260 \text{ md}$$

The measured permeability is the average of the two halves.

In case B, the lateral variation, the two halves of the plug must maintain the same air flow but would require different pressure differentials to do so. The total pressure differential would be the sum of these two:

$$\Delta P = \Delta P_1 + \Delta P_2$$

$$= \frac{Q \mu \frac{L}{2}}{AK_1} + \frac{Q \mu \frac{L}{2}}{AK_2}$$

$$= \frac{Q \mu L}{2A} \left[\frac{1}{K_1} + \frac{1}{K_2} \right] \tag{1-14}$$

Substituting this pressure differential into equation (1-9) we may determine the measured permeability:

$$\text{Whole Plug } K_m = \frac{Q \mu L}{A \, \Delta P} \tag{1-9}$$

$$= \frac{Q \mu L}{A \cdot \frac{Q \mu L}{2A} \left[\frac{1}{K_1} + \frac{1}{K_2} \right]} \tag{1-15}$$

$$= \frac{2}{\left[\frac{1}{K_1} + \frac{1}{K_2} \right]}$$

$$K_m = \frac{2}{\left[\frac{1}{500} + \frac{1}{20} \right]} = 38 \text{ md}$$

The measured permeability is more strongly influenced by the lower permeability fraction. This is true even if only a small portion of the plug has the lower permeability. If, in this example, the lower porosity and permeability portion occupied only one tenth of the length of the core, the measured porosity of the whole core would be 17 percent, while the measured permeability would be 147 md.

Depending upon size and orientation, low permeability inclusions in a core plug may drastically reduce the measured permeability. Such inclusions may result from clay swelling or mud solids invasion. They may represent natural inhomogeneity in the formation.

Care in selecting clean, undamaged, and homogeneous samples for permeability measurements is essential. Even then, the measured permeability for the plug can be only a gross estimator of flow behavior in the formation as a whole.

Finally, there are theoretical considerations which affect the reliability of permeability estimates. The d'Arcy equation, equation (1-9), requires certain physical conditions in order to hold true. These conditions are only partially met in the permeameter measurement. Flow in the formation occurs under conditions which differ not only from the theoretical condition but also from those in the permeameter. Figure 1-21 gives a comparison of the three sets of conditions.

COMPARISON OF THEORETICAL AND ACTUAL FLOW CONDITIONS

Theoretical (d'Arcy)	Ruska Permeameter	Reservoir
Linear Flow	True	Radial Flow
Horizontal Flow	Vertical Flow	Possibly Horizontal and Vertical Components
Laminar Flow	Laminar Flow	?
Incompressible Fluid	Compressible	Compressible
Isothermal Conditions	Approximately True	Approximately True in Short Term
Steady State Flow	Approximately True	Possibly Discontinuous Flow
Medium is totally saturated with a single constant viscosity fluid	Approximately True	Medium will be saturated with multiple flowing phases, discontinuous and immobile phases of varying viscosity

Figure 1-21. Comparison of Theoretical and Actual Flow Conditions

The Ruska Permeameter measures the absolute permeability

- of a small rock sample
- in only one orientation
- to air
- in linear steady-state flow
- under conditions which approximate to the ideal

These factors must be considered in the evaluation and use of permeability estimates.

1.24 SATURATION

1.25 Measurement

The standard instrument for determining saturations is the Saturation Still (Figure 1-22). It operates in a similar manner to the Mud Engineer's Mud Still. A weighed sample of fresh core fragments is heated in a sealed retort. Evolved vapors are condensed in a water cooled condenser and collected as oil and water in a calibrated collection tube. Using the known porosity of the rock, from the porometer test, it is possible to calculate the saturations of oil and water in the sample:

$$V_{ps} = V_p * \frac{W_s}{W_p} \qquad (1\text{-}16)$$

where

V_{ps} = pore volume of the saturation sample, cc

V_p = pore volume of the porosity sample, cc

W_s = weight of the saturation sample, gm

W_p = weight of the porosity sample (before fluid extraction and drying), gm

$$S_w = \frac{V_w}{V_{ps}} \qquad (1\text{-}17)$$

$$S_o = \frac{V_o}{V_{ps}} \qquad (1\text{-}18)$$

where

S_w =   water saturation, fractional (0-1.0)

S_o =   oil saturation, fractional (0-1.0)

V_w =   recovered volume of water, cc

V_o =   recovered volume of oil, cc

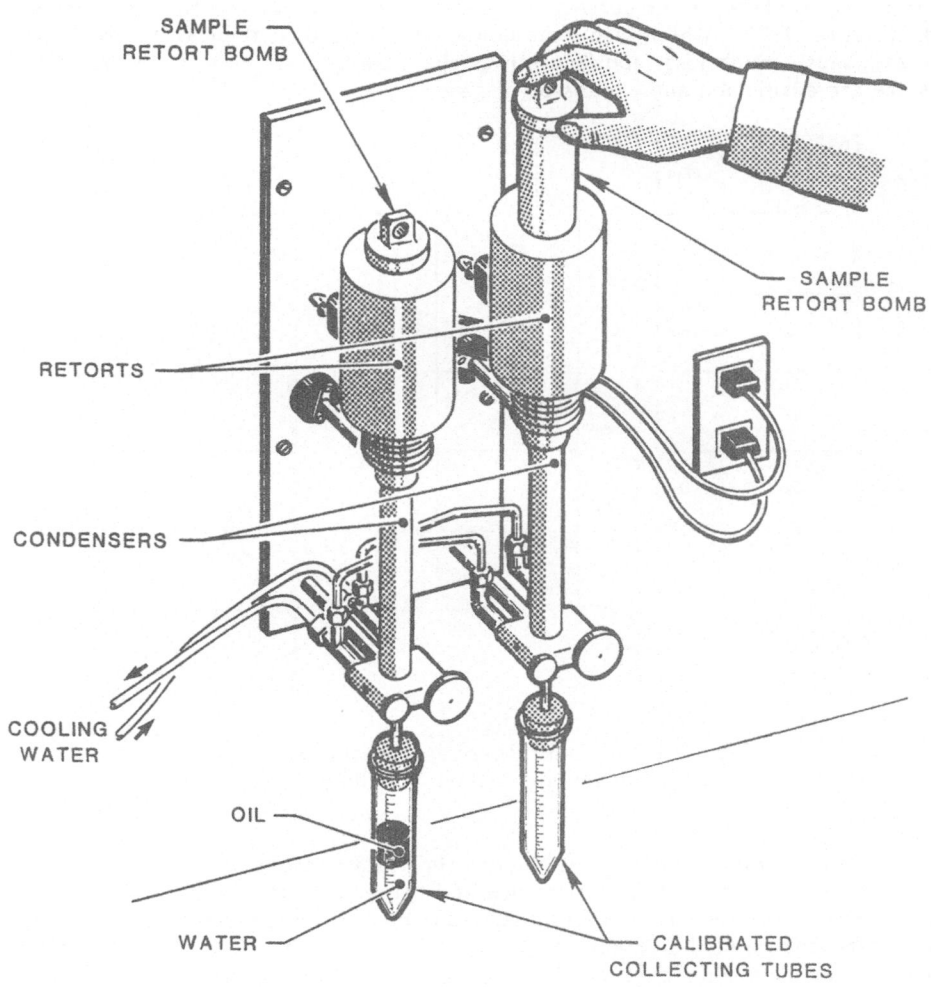

Figure 1-22. Saturation Still

1.26 Limitations

Saturation determinations may be marginally affected by the porosity-modifying mechanisms discussed above. The most important effect on measured saturations is the mobility of the fluids themselves. During drilling, mud filtrate is flushed into the formation ahead of the core bit, expelling much of the original formation fluid. During recovery to surface, and at surface, reduction in pressure and temperature results in exsolution of dissolved gas from the water and oil phases, and shrinkage (or volume reduction) in those phases. Expansion of evolved gas results in a volume increase in that phase and some expulsion of oil and water and further reduction in volume in those phases.

Figures 1-23, -24 and -25 are representations of the types of modification of reservoir fluid saturations, in place, after cutting and after retrieval. Notice that oil-based drilling fluids (Figure 1-25), which are commonly less invading than water-based fluids (Figure 1-24), cause much less modification as the core is cut. In both cases, the major modifiers are the exsolution and expansion of gas.

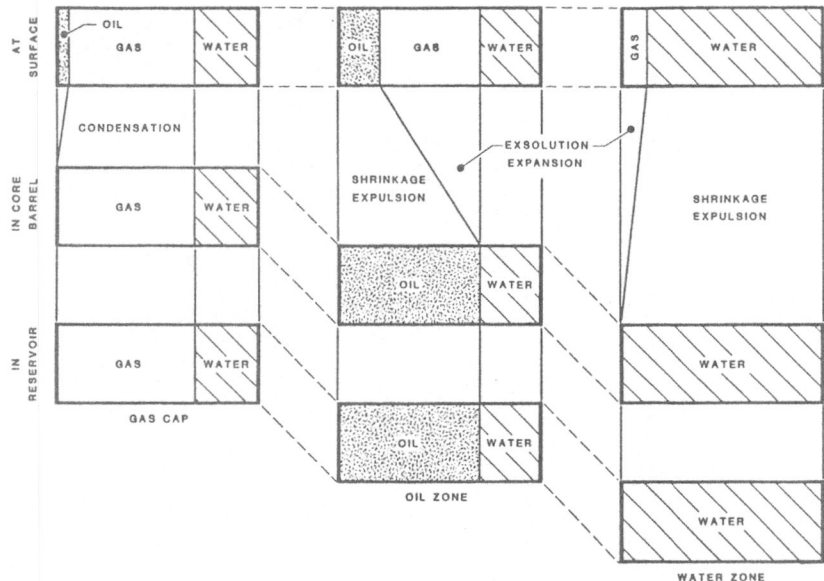

Figure 1-23. Fluid Recovery, Water-Based
Drilling Fluid without Flushing

The coring procedure will alter to some degree the fluid content of the core. Where the borehole pressure exceeds formation pressure and there is permeability, the mud filtrate will displace the formation fluid in the core sample by an amount depending on the degree of overbalance and the effective porosity and permeability of the reservoir rock. The flushing of a permeable core during cutting occurs even with low-water-loss muds since filter cake is not allowed to build up on the fresh surfaces continually exposed by the bit. A flushed core will therefore show less oil and gas, and more water saturation, than an unflushed core.

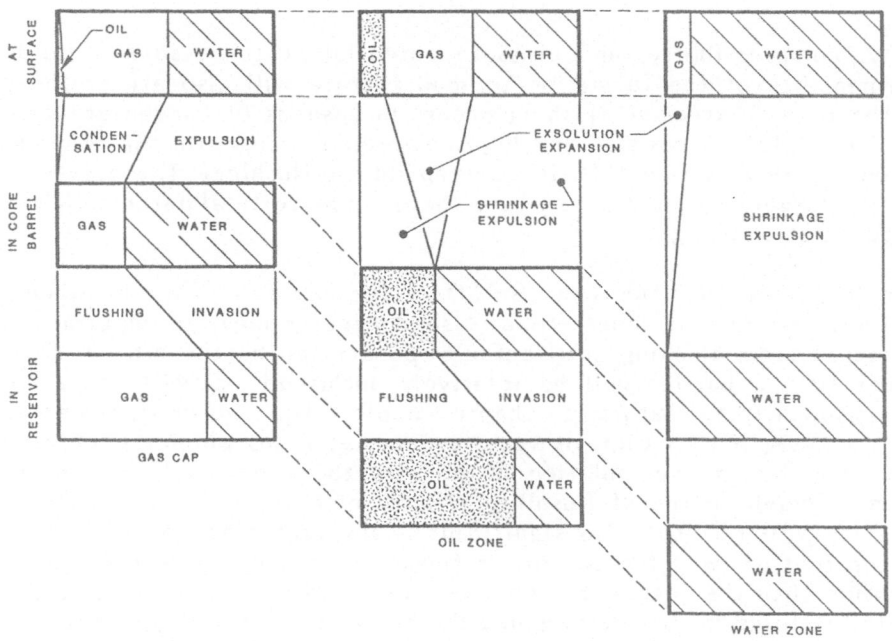

Figure 1-24. Fluid Recovery, Water-Based Drilling Fluid with Flushing

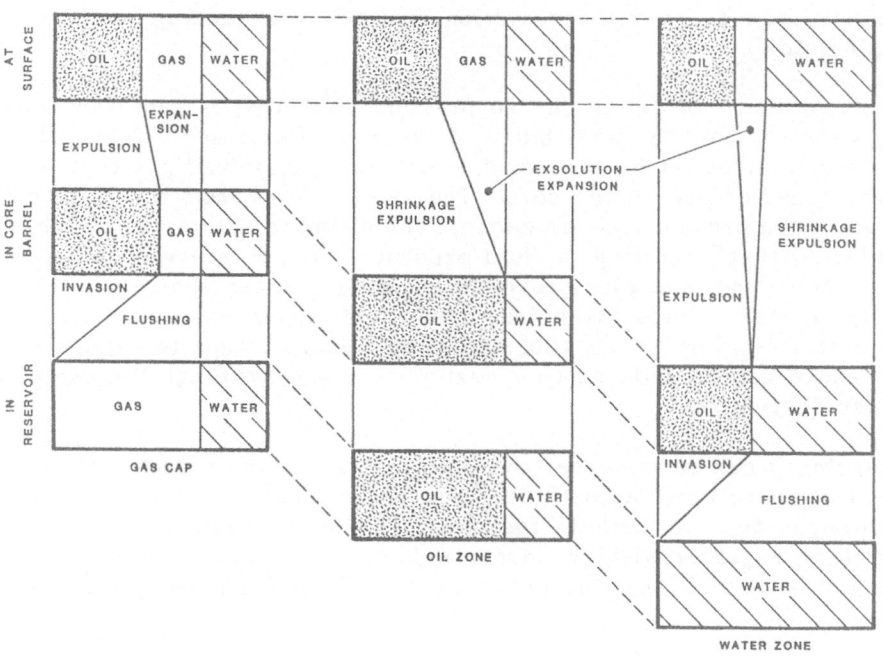

Figure 1-25. Fluid Recovery, Oil-Based Drilling Fluid

The viscosity, compressibility, and pressure potential of the reservoir fluids and the relative permeability of the formation to mud filtrate will also affect the degree of flushing in the core. Extensive flushing occurs in gaseous or condensate zones due to these fluids having low viscosities and high compressibilities. On the other hand, low-gravity viscous oil reservoirs will be less susceptible to flushing. The degree of flushing will be less in a larger diameter core where there is low vertical permeability and where the rate of penetration is high.

A certain stratigraphic condition can also affect the degree of flushing when combined with a relatively fast rate of penetration. A sand directly under an impermeable horizon will be protected from flushing until actually penetrated by the bit. A sample taken directly under such a barrier will be relatively unflushed and will contain higher oil saturation and lower water saturation than a sample taken two or more feet below the barrier. It is also significant that, in water-productive sands which do contain some oil, there will usually be an accumulation of oil directly under an impermeable barrier. Coupled with a lesser degree of flushing, a fairly high residual oil saturation and low water saturation will result. Also significant is the fact that, in permeable oil sand, usually the top foot or two of each core is flushed to a lower oil saturation and higher water saturation than the rest of the core because mud is usually circulated above the core to obtain "bottoms up" before dropping the ball and commencing coring.

As the partially or completely flushed core is recovered to surface, its pressure and temperature are reduced to atmospheric conditions. The gas dissolved in the oil and water expands and subjects the core to a solution-gas drive. The expulsion of liquids continues as the core nears the surface until the relative permeability to gas within the core approaches 100 percent. No further expulsion occurs at this point, but evolution of gas and possibly some retrograde condensation may occur, depending on the composition of the reservoir fluids.

The amount of expulsion of fluids due to pressure reduction is generally governed by permeability, viscosity and compressibility of reservoir fluids and original fluid saturation, and is directly affected by the amount of free or dissolved gas that was initially present and the speed of pulling the core. The loss of fluids from a core sample by the expansion of gas has a pronounced influence on the ultimate saturation of the core. The action is similar to that occurring in fluid expulsion from a reservoir by gas expansion except that, in this case, the pressure differential is greater and the loss of fluids is correspondingly greater. Cores taken from deep sands experience more pressure reduction than cores from shallow sands, thus effecting a more complete expulsion. Loss of fluids by this cause can be reduced to a negligible amount through the use of pressure core barrels (see Section 2).

From the foregoing it can be seen that the fluid saturations measured at the surface will be markedly different to those actually in place in the reservoir. But from the information gained through the core analysis tests, gas evaluation while drilling, wireline logs and drillstem tests, together with an understanding of the processes at work, a reasonably accurate picture of the reservoir and its contained fluids can be formulated.

2

CORING PROCEDURES

2.1   GENERAL

Several different methods and tools are available for obtaining core samples. The decision on which to use is made by the Oil Company engineers and geologists on the basis of a number of criteria. The most important of these are

- What types of data are required?
- Over what depth interval is data required?

On the basis of these criteria, a decision will be made as to which of the two basic methods of coring to adopt

- Bottomhole Coring
- Sidewall Coring

Bottomhole coring obtains large diameter (3 to 5 inches) and length (30 to 90 feet) cores of relatively undisturbed formation. However, bottomhole coring proceeds much less rapidly than conventional drilling and involves expensive diamond bits and core barrels. Rig time and tool rental costs make bottomhole coring an expensive operation. The decision to core must be made prior to entering the zone of interest. Therefore, an erroneous depth correlation may result in unnecessary coring of overlying formations.

Sidewall coring obtains small (1 inch by 2- or 3-inches) cores of intact though generally contaminated or damaged formation. They are cut from the borehole wall after drilling, using a tool run on a wireline. Sidewall cores may therefore be obtained quickly and relatively cheaply over extensive depth ranges. Selection of sidewall core points is made after drilling and logging the well, eliminating the need for unnecessary cores.

On review of data range, required sample quality and cost factors, a method of coring is selected. Variations of the two basic methods are then selected in consideration of drilling requirements and formation characteristics.

2.2   BOTTOMHOLE CORING

As the name implies, this involves the cutting of a core at the bottom of the hole. In most cases it involves the drilling of hole with a hollow bit, allowing a solid cylinder of uncut formation to enter an inner retainer or core barrel which may later be retrieved to surface. (There is one exception to this, discussed in Paragraph 2.15.)

2.3   CONVENTIONAL CORING

The drilling assembly for conventional bottomhole coring consists of two parts: the core head or bit and the core barrel or retainer. Figure 2-1 shows the complete assembly and

34

its components. Above the coring assembly a conventional bottomhole assembly of drill collars and stabilizers is run.

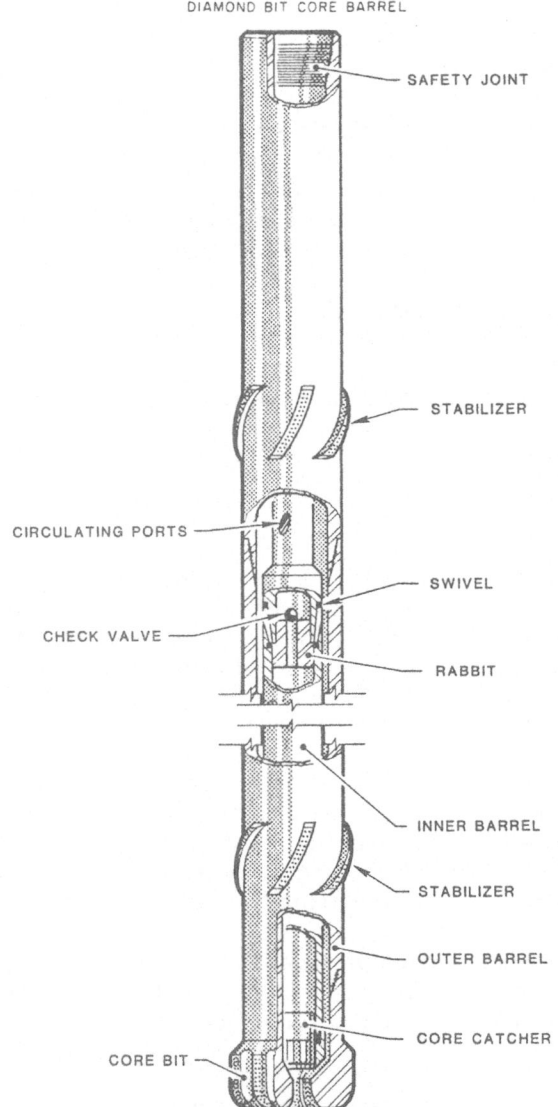

Figure 2-1. Conventional Core Barrel

2.4   The Core Bit

Core bits are similar to conventional drill bits in design and function. Unlike drill bits, however, they do not cut a complete cylinder of formation. Instead, they are designed to cut an annular ring or kerf of formation, leaving a solid cylinder of uncut formation to pass through the center of the bit into the core barrel above.

Like drill bits, core bits are available in various mechanical designs. Drag-type and rotary cutter-type core bits are available and are used in coring soft to medium formations. However, diamond core bits which offer long life and the ability to cut a range of formation types have become the predominant type of core bit in use (see Figure 2-2).

A) DRAG CORE BIT

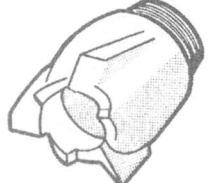

B) ROTARY CUTTER CORE BIT

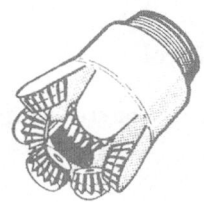

C) DIAMOND CORE BIT

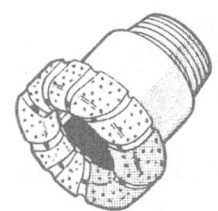

Figure 2-2. Core Bits

The diamond core bit differs from the diamond drill bit in having two circulation systems. When first run into the hole, circulation of drilling fluid passes through the center of the bit in the conventional manner. When coring, fluid may be diverted through discharge ports in the face of the bit (see Paragraph 2.7).

Because of the high cost and comparatively small use of diamond core bits, a drilling operation is rarely equipped with a supply in all types and sizes. Commonly only two types, unconsolidated/soft and medium-medium hard formation bits, and one size, the smallest expected hole size, are kept available.

Another reason for keeping only one size of core bit on hand is the strong relationship between the success rate in core recovery and the ratio of the size of core to the size of the hole. The larger the kerf, the more likely the core will be damaged by vibration and mud circulation erosion, and therefore the poorer the likely recovery.

To ensure good recovery it is necessary to either have available a range of core barrels to give large core-to-hole size ratios for all hole sizes or to choose a single core size and limit the range of bit sizes with which it is used. Economy and standardization dictate the latter solution, and most drilling operations have core bits and core barrels of only a single size, most commonly an 8-1/2-inch bit which cuts a 3-1/2-inch core. If it is decided to core in a larger size hole, a small size bit is used, leaving an undersize hole (a 'rat hole') which must later be reamed.

Diamond and conventional core bits are easily damaged by junk or shocks. Operating procedures like those for a diamond bit should be used. The hole must be free of junk. Running in the hole should be done slowly to avoid damaging or plugging the bit by hitting a bridge or dogleg. Initial weight-on-bit and rotary speed should be very low and increased slowly up to the operating levels which should then be held constant. Pump pressure and flowrate should be held low to prevent the bit from being pumped off-bottom and bounced.

2.5   The Core Barrel

A conventional core barrel (Figure 2-1) consists of two parts: an inner and an outer barrel.

2.6   The Outer Barrel provides strength to the complete assembly and connection to the rotating drillstring. It is normally thirty feet long but up to three sections may be joined, allowing up to ninety feet of core to be cut.

At its lower end the outer barrel is attached to the core bit and at its upper end to the bottomhole assembly. Attachment to the BHA is via a safety joint which allows the drillstring to be backed-off and removed, should the core barrel become stuck in the hole.

Two bit-size stabilizers are attached to the body of each section of outer barrel. These are essential in maintaining a straight hole while coring. Due to the generally lower weight-on-bit used in coring, there is a tendency for drillstring 'wobble' which may result in tilting or even bending of the core barrel. This will cause the core to jam, preventing further coring and possibly permanently damage the core barrel. The extra stabilization prevents this.

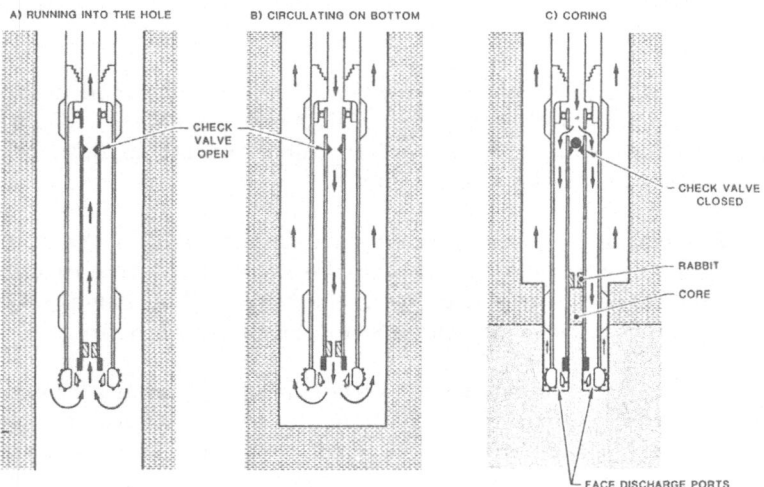

Figure 2-3. Operation of a Conventional Core Barrel

2.7   The Inner Barrel is a much thinner walled tube, the purpose which is to hold the core for retrieval to surface. Like the outer barrel it can consist of up to three, 30-ft-long sections. Inside diameter, and hence core size, ranges from three to five inches with 3-1/2 inches being the most common. It is attached to the outer barrel by a ball bearing swivel allowing rotation but not vertical movement. Thus the inner barrel is carried in and out of the hole supported by the outer barrel but is free to remain stationary relative to the formation while the outer barrel and core bit rotate.

At the upper end of the inner barrel is a check valve and circulation ports. When the valve is open, fluid flow through the drillstring passes through the inner barrel and the center of the core bit (see Figure 2-3a and b). When the valve is closed, by dropping a steel ball down through the string, mudflow is diverted out through the circulation ports between the inner and outer barrel and through the discharge ports in the bit face (Figure 2-3c). Drilling fluid displaced from the inner barrel as the core enters is free to pass upward through the check valve.

At the lower end of the inner barrel is the core catcher. This is a spring-loaded latch or a set of small slips which prevents the core from sliding out of the barrel when the drillstring is lifted. Inside the inner barrel is the rabbit. This is a metal plug free to move through the inner barrel with a close fit to the inside diameter. A hole through its center allows fluid to escape. As coring progresses and core enters the inner barrel, the core pushes the rabbit upward, wiping the barrel inside surface and removing the debris which may cause the core to jam (Figure 2-3). When the core is removed from the inner barrel at surface, by removing the core catcher and allowing it to slide out, the rabbit follows the core out of the barrel removing all fragments. Free movement of the rabbit through the inner barrel at surface also provides a check that the barrel has not been bent or distorted.

2.8   Coring

After the inner and outer core barrels and the core bit have been assembled, they are run into the hole as described above. Pup joints of drillpipe are added to the drillstring so that, when the core bit reaches bottom, the kelly can be attached with as much 'kelly-up' as possible. This allows the maximum length of core to be cut before a connection is required.

When the bit is a short distance above bottom, the kelly is attached and circulation begun through the inner barrel. Pump pressure is carefully monitored. Abnormally high pump pressure indicates that there may be debris in the barrel or core catcher which must be pumped clear before coring can commence. When pump pressure is stable the circulation is halted, the kelly broken-off, and a steel ball dropped into the drillstring. The kelly is reconnected, and circulation continued until a rise in pump pressure indicates that the ball has seated in the check valve and flow is being diverted through the circulating port between the inner and outer barrels and out of the discharge ports in the bit face. Pump pressure is again monitored to check that the discharge ports and the bit face are clear of debris. Coring then begins.

Drilling parameters (weight-on-bit, rotary speed, pump pressure and flowrate) are monitored and maintained as constant as possible. Any changes are made slowly and carefully. Sudden changes in drilling parameters may damage the core bit or cause the core to break. Broken core fragments may lodge in the core catcher, jamming it and bringing coring to a halt. Core jamming occurs most commonly in brittle or fractured formations and when connections are made. At a connection, the core is broken while lifting the string.

When coring it is important to maintain optimum drilling parameters and rate of penetration. In addition to economic and bit wear considerations, there is also an effect on the success in core recovery. If excess weight-on-bit or rotary speed is applied in an attempt to increase rate of penetration, the core may be broken or burned. On the other hand, if, rate of penetration is low, the time between the core being cut and its entering the inner barrel will be long. During this period it is subject to washing by the circulating drilling fluid. The core may be fragmented by this or may have its diameter reduced so that it cannot be securely held by the core catcher. The distance between the core bit and the core catcher can be varied to help alleviate this problem. For unconsolidated formations, which are most susceptible to washing, the core catcher is set as close as possible to the core bit. In harder formations, less affected by washing but more likely to break, the distance is increased to reduce the risk of broken fragments under the bit being thrown up and jamming the core catcher.

Sudden increases in pump pressure may indicate the bit is plugged. This may be cured by lifting the bit off-bottom and continuing to circulate. If not, the bit must be pulled since further coring would cause it to become overheated and damaged.

When the core barrel is filled, or jammed, rate of penetration rapidly decreases. The bit is allowed to drill-off the weight-on-bit at a slightly increased rotary speed. This ensures a clean cut at the bottom of the core. The bit is slowly picked up off-bottom with circulation continuing.

When picking up off-bottom, either with a full core barrel or in order to make a connection, the brake is operated gently and the weight indicator carefully watched. If the core catcher operates correctly, an overpull will be observed first. When the core breaks, weight will fall back to normal. Overpull should not be allowed to exceed 30,000 lb above string weight or the maximum rating of the safety joint (whichever is least) as this may damage the core barrel or cause the string to back-off from it. If the core cannot be broken, the brake is set with 20,000 to 30,000 lbs overpull, and the core is rocked with small forward and back movements of the rotary table until the core breaks.

The core catcher should latch and break the core at bottom. However, some slippage or delay in latching may result in the loss of a small section of the bottom of the core. If this occurs, the string should be picked up at least 20 feet off-bottom and then lowered slowly back to bottom with the weight indicator being carefully monitored. When approximately 500 lbs of weight is seen, the rotary is turned slowly and intermittently. The barrel should slide slowly over the core with further increase in weight-on-bit. When the core fragment is safely in the barrel, pickup can be attempted again or, if after a connection, coring can recommence. If the core fragment cannot be guided into the barrel or if it cannot be held on a second attempt, it is likely that the fragment has fallen to the side of the hole or has fractured into small pieces. In either case, it should be abandoned. Further attempts at retrieval may jar and damage the bit or overwork the core catcher, causing further core to be lost from the barrel.

Although the core catcher normally works reliably, there is a possibility, especially with broken or unconsolidated cores, that circulation, shock or vibration may loosen the core and cause it to fall from the barrel. It is therefore common practice to pull out of the hole carefully without circulating bottoms-up. Pipe is pulled slowly and stands broken-out with a spinning chain on the upper connection rather than by turning the string in the rotary table. Particular care is taken pulling out of the reduced diameter rat-hole, as the large diameter outer barrel and stabilizers can, if moved too quickly, produce sufficient swab pressure reduction to literally suck the core out of the barrel.

At surface, the outer barrel is hung in the slips. The inner barrel is lifted out of it. With the inner barrel hanging vertically in the derrick, the core catcher is removed and the core allowed to slide out. If the core is seriously jammed, if it is suspected to contain hydrogen sulfide, or if the formation is so unconsolidated that falling from the barrel is likely to disrupt the core, the inner barrel may be laid down on the catwalk and the core pumped out of the barrel by attaching a high-pressure air, water or mud line at the upper end. Retrieval of cores at surface is discussed in much greater detail in Section 3.

When failure in pick-up or poor recovery of a well-consolidated core suggests that core pieces have been left on bottom, great care should be taken on the next core bit run to avoid bit damage or jamming the core catcher. Washing over the core fragment as described above should first be attempted. If this is not successful, recoring over the fragments should be achieved slowly with low, steady weight-on-bit until the previous total depth is reached. If the formation is particularly hard and brittle it may be desirable to run a drill bit and junk sub to clean out the hole bottom before taking subsequent cores.

2.9   RETRIEVABLE CORE BARRELS

Retrievable, or retractable, core barrels are similar in design and operation to conventional core barrels. They have the added feature that the inner barrel can be run-in and pulled-out of the drillstring without tripping. Thus, when the inner barrel is filled, it can be removed and replaced in order to allow continuous coring over long intervals.

In order to be retrievable, the inner barrel must be slim enough to pass through the tool joints of the drillstring. The cores obtained are, therefore, smaller than cores from an equivalent sized conventional core barrel, commonly two to three inches in diameter. This also results in a larger kerf and less successful recovery rates (see Paragraph 2.4).

Since continuous coring is rarely performed in petroleum exploration and since core size and recovery are important for core analysis, retrievable core barrels are not often seen. Their use is most common when the purpose of the core is geological rather than petrophysical (for example in mineral exploration). They are also widely used in deep ocean exploration programs where the lack of a marine riser prevents cutting recovery and makes tripping and hole reentry difficult.

Several types of retrievable core barrel are available, but they can be classified on the basis of the two methods of retrieval.

2.10 Wireline

There are two types of wireline retrievable core barrels. In both, the inner barrel may be retrieved with an overshot run on a wireline. A new inner barrel may then be run in and located in the outer barrel.

The most common type is shown in Figure 2-4a. The retrievable inner barrel is in all ways similar to that of a conventional core barrel except for the retriever connection at its upper end.

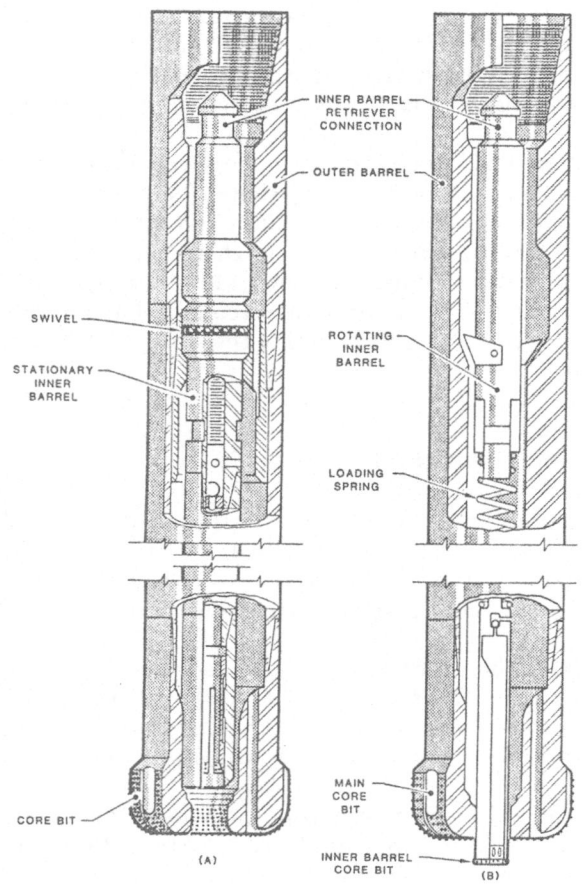

INNER BARREL
RETRIEVER
CONNECTION

OUTER BARREL

SWIVEL

ROTATING
INNER
BARREL

STATIONARY
INNER
BARREL

LOADING
SPRING

MAIN
CORE
BIT

CORE BIT

(A)

INNER BARREL
CORE BIT

(B)

Figure 2-4. Wireline-Retrievable Core Barrels

The second type of wireline retrievable core barrel is unlike the previous cases in that the inner barrel, when installed, is locked to the outer barrel and rotates with it (Figure 2-4b). Below the core catcher is a small core bit which protrudes through and a few inches ahead of the main core bit. The inner barrel is spring loaded so that, should a hard formation be encountered which may damage the small bit, it is pushed back inside the protection of the main bit.

The advantage of this tool is that, on each run, part of the cutting structure is replaced to maintain good coring rates. The small core bit cuts a very small kerf, giving a large core-to-hole-size ratio and better core recovery. The main core bit following behind acts as a reamer opening the hole to full guage.

2.11 Reverse Circulation

This is similar to the conventional core barrel but the inner barrel is not mechanically attached to the outer barrel. It is held in place by centralizers and pump pressure. When the core barrel is full, the hole is reverse circulated and the complete inner barrel is pumped up to surface through the drillstring.

2.12 RUBBER SLEEVE CORE BARREL

When coring unconsolidated lithologies, it is common to obtain relatively good recovery to surface but for the core to suffer serious damage or disruption during removal from the barrel and later handling. The rubber sleeve core barrel offers a solution to this problem by enclosing the core in a shrink-fit rubber tube as it enters the inner barrel. The complete core enclosed in the tube can be removed from the barrel without disruption and cut into convenient length for core analysis. For later visual inspection of the core, it can be artificially consolidated by freezing or by injecting a plastic gel.

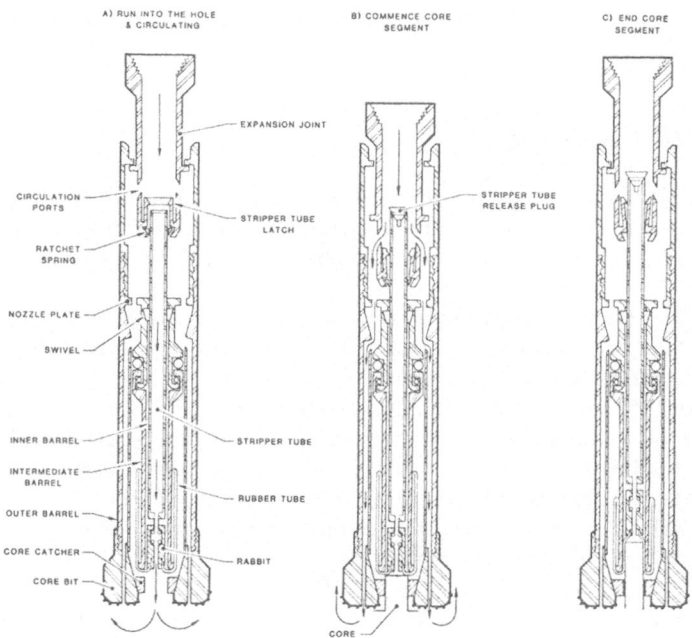

Figure 2-5. Rubber Sleeve Core Barrel

The rubber sleeve core barrel consists of four concentric tubes -- the outer, intermediate and inner barrels and the stripper tube (Figure 2-5a). When running into the hole and circulating mud, flow is through the stripper tube and central orifice of the core bit.

Before coring commences, a dart-shaped release plug is dropped. This seals the top of the stripper tube diverting mudflow through circulation ports, between the intermediate and outer barrels and through the bit face discharge ports. The plug also releases the stripper tube latch so that when weight is applied the expansion joint closes (Figure 2-5b). A ratchet spring grips the stripper tube so that when the expansion joint closes it slides down over the stripper tube, but when the joint opens it grips and lifts the stripper tube.

Circulation and rotation begin as normal but no weight is applied to the bit (beyond that required to close the expansion joint). Force on the bit is supplied by the hydraulic impact of drilling fluid passing through nozzles in the nozzle plate (Figure 2-5a). The number and sizes of nozzles can be selected to give appropriate force for the formation to be drilled.

As coring progresses, core enters the barrel, the lower part of the barrel moves downwards and the expansion joint progressively opens. Opening the expansion joint pulls the stripper tube up through the barrel which pulls rubber tube into the barrel around the core as it enters.

After coring two feet, the expansion joint will be fully open and coring will stop (Figure 2-5c). Weight is applied to reclose the expansion joint, pushing it further down over the stripper tube, and another 2-ft segment is cored. Coring continues in 2-ft segments until the barrel is full. Circulation is then stopped and the core recovered normally to surface.

2.13 ORIENTED CORE BARREL

One of the geological advantages of cores over cutting samples is that large-scale sedimentary and diagenetic structures may be identified. The dip of beds, fractures and other sedimentary or diagenetic structures in the core may be recognized and estimated.

In a conventional core barrel, such estimates are possible with an accuracy controlled by the inclination of the cored hole. For example, if a structure has a dip of 30° relative to the core and the borehole has an inclination of 3°, then the true dip of the structure may be between 27° and 33°. Since the horizontal orientation of the barrel is unknown, strike directions and true dips cannot be estimated. Even relative strike directions and true dips of structures in different parts of the core may not be known if the core is broken between those points. Rotation of the inner barrel when off-bottom and rotation of the core during retrieval prevents accurate orientation of the broken core pieces unless perfect fit at the broken ends can be made.

Some types of core barrels are capable of being field-modified to allow horizontal and vertical orientation of the core as it is cut. This involves installing a multishot survey tool in a nonmagnetic drill collar immediately above the core barrel. The survey tool is fixed to the inner barrel, allowing it to remain stationary with the barrel when coring or move with it if rotation occurs off-bottom. Continuous records of the hole inclination, azimuth and tool face (orientation of the inner barrel) are made while coring.

Orientation of the core within the barrel and after recovery is made by an 'orienting shoe' attached to the core catcher at the lower end of the inner barrel. The orienting shoe contains three knives which scribe reference grooves on the core as it enters the inner barrel.

Combining the timed multishot survey measurements, rate of penetration and core orientation marks, it is possible to accurately orient the whole core length and provide accurate dip and strike of structures within the core. Core orientation also allows the preparation of oriented samples for core analysis and for mineralogical and rock mechanical investigation.

2.14 PRESSURE CORE BARREL

As discussed in Paragraphs 1.17 through 1.26, the major changes in core properties from those in-situ occur during recovery to surface. Decline in confining pressure and, to a lesser extent, temperature result in the relief of rock stresses and the modification of absolute and effective porosity and permeability. The exsolution and expansion of gas substantially modify the relative saturations of reservoir fluids.

In the past such limitations were accepted, and the combination of core analysis with wireline and DST data was used to produce acceptable estimates of reservoir characteristics and performance. In recent years, with increased drilling of deeper, less porous reservoirs and emphasis on reservoir stimulation, secondary and tertiary production, a greater need has arisen for accurate knowledge of porosity, permeability and saturations in situ. The pressure core barrel and associated pressure coring techniques have been developed to obtain cores which, as closely as possible, retain a composition and properties representative of the reservoir when recovered to surface.

The pressure core barrel is similar to a conventional core barrel but has an upper O-ring pressure seal and lower ball valve pressure seal allowing the complete inner barrel to be enclosed within the lower part of the outer barrel and maintained at formation pressure during recovery (see Figure 2-6). To compensate for pressure changes due to cooling in retrieval, pressure-regulated nitrogen is allowed to enter the barrel from a nitrogen reservoir.

The pressure core barrel uses a conventional diamond or synthetic "stratapax" core bit. Up to nine feet of 2-1/2-inch diameter core may be cut in a single coring operation. The smaller-than-normal core size is offset by the considerably better core quality.

Although the pressure core barrel reduces the modifying effects of core recovery, the core is still subject to the lesser but significant effect of flushing during drilling. There is an attempt to reduce flushing of the core to a minimum by careful control of mud density, fluid loss and flowrate. This involves the Logging Geologist or GEMDAS Operator in providing reliable estimates of Formation Balance Gradient and in running Nitrate Ion or other tracer tests.

The pressure core barrel is made up on surface as shown in Figure 2-6a. The pressure regulator is set to the estimated formation pressure and the nitrogen reservoir is charged to approximately twice this pressure. The complete assembly is then leak-tested by immersion in water for thirty minutes. When confirmed pressure-tight, the core barrel is made up in the drillstring and run to bottom normally. To avoid swab or surge damage to the pressure seals, a pressure core barrel is always run with a core bit which is smaller than the current open hole diameter.

GEMDAS® is an Exlog registered service mark, standing for Geological and Engineering Monitoring and Data Analysis Service.

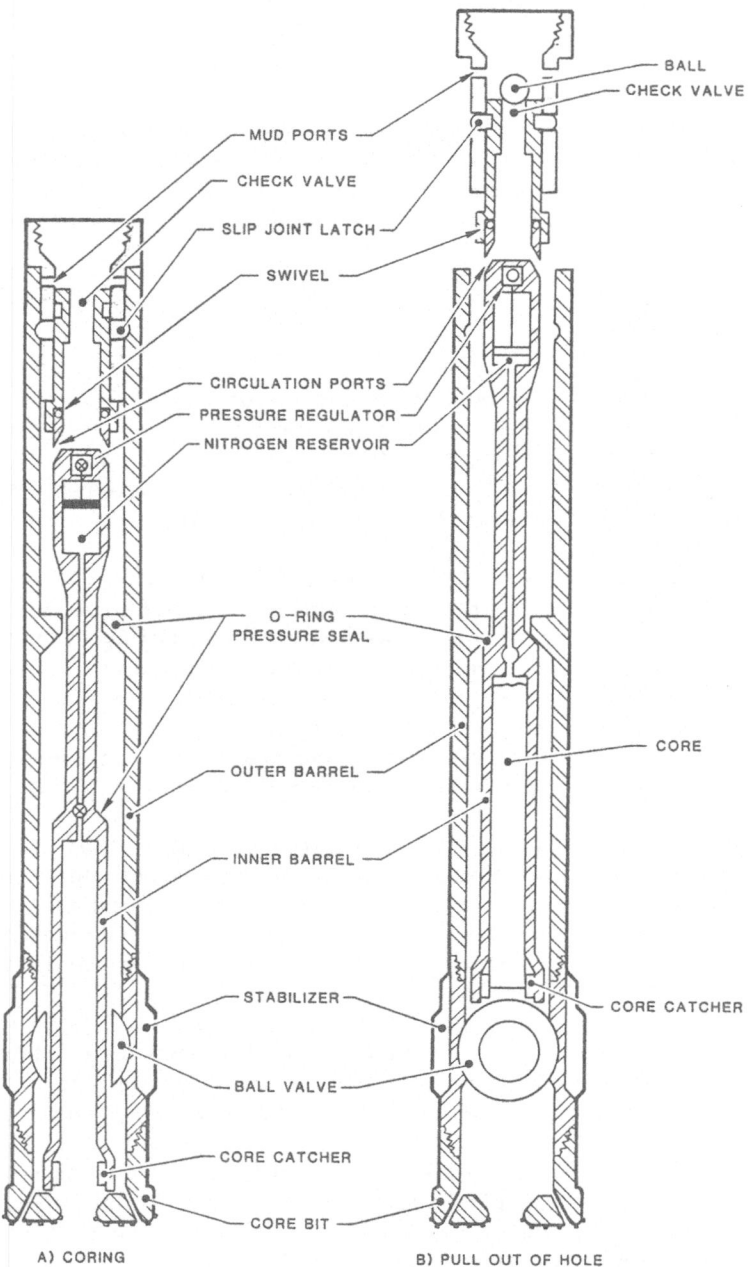

Figure 2-6. Pressure Core Barrel

On bottom, circulation is commenced at a high flowrate. Drilling fluid passes through the circulation ports, between the inner and outer barrels and through the discharge ports in the bit face. The hole is displaced with polymer stabilized drilling fluid designed to minimize flushing. Desirable properties are

● Mud density exceeds Formation Balance Gradient by the minimum acceptable safety margin

- Moderate to high viscosity

- Ultra-low fluid loss

- High solids content of particle size sufficient to bridge minimum pore opening (one-third opening diameter or larger)

- No emulsifiers, dispersants, oils or lubricants

- Stabilized concentration of Nitrate Ion or Tritium Ion tracer

In addition, yield point, gel strength and surface mud treatment must be sufficient to maintain good cuttings recovery and stable mud properties.

When circulation is stable, coring is commenced using weight-on-bit and rotary speed as in conventional coring but very low flowrate to reduce flushing. Core enters the inner barrel and is retained by the core catcher as normal. When the core barrel is full, circulation is continued as the core barrel is pulled out of the rat hole into the full-gauge open hole section. Circulation is then stopped and the kelly broken out. Mud samples are taken at the suction line and the flowline. These are analyzed for tracer concentration for comparison with later analyses of core pore waters in order to determine the degree of mud filtrate invasion.

A 1-1/4-inch steel ball is dropped into the top of the string, the kelly is reconnected and the ball pumped down to the core barrel. When the ball seats in the check valve, a pump pressure surge occurs, forcing down a piston and opening the slip joint latch. The slip joint opens, allowing the outer barrel to fall onto and seal the o-ring pressure seal, closing off the mid-point of the outer barrel. Simultaneously, the ball valve passes down over the core catcher and rotates through 90° to close and seal off the lower end of the outer barrel (Figure 2-6b). Closure of the pressure seals also activates the pressure regulator which maintains pressure in the enclosed volume by release of nitrogen from the reservoir at regulated pressure.

When the outer barrel has fallen and the seals have closed, the mud ports above the check valve are opened to the annulus. Successful accomplishment of closure is therefore indicated by a fall in pump pressure. The pump may be stopped and the core barrel retrieved from the hole normally.

At surface, the complete lower barrel assembly is removed and placed in dry ice for several hours to freeze the contents. After freezing, the nitrogen pressure is released and the inner barrel removed. The inner barrel and contained core is then cut into sections of a convenient size for packing, the section ends sealed and the core shipped to the core laboratory. Procedures for handling and sampling frozen cores will be discussed further in Section 3. Core analysis on frozen core material is rarely, if ever, performed at the wellsite and is not covered in this manual.

2.15 CORE EJECTOR BIT

When drilling with a diamond drill bit, it is commonly difficult to maintain good stratigraphic correlation because of the poor quality of pulverized or burnt drill cuttings. Where larger samples are required for lithological evaluation a diamond core ejector or crusher may be used.

46

The core ejector bit is a hybrid between a regular diamond drill bit and a core bit (see Figure 2-7a). The face of the bit has three circulation ports surrounding a hole through which a small core can enter the bit. There is no internal core barrel or catcher, and, after passing through the bit, the core is usually broken by drilling fluid turbulence or a core crusher sub. The core crusher sub has a cam-shaped profile which impacts and breaks the core into small recoverable fragments (Figure 2-7b).

When the bit is picked up off-bottom, core fragments may fall out of the bit and be circulated to surface. Alternatively they will remain inside the bit being carried back to surface and recovered when the bit is tripped. Because of the imprecise knowledge of their depth of origin and long immersion in drilling fluid, ejected cores provide incomplete lithological information and are unsuitable for core analysis.

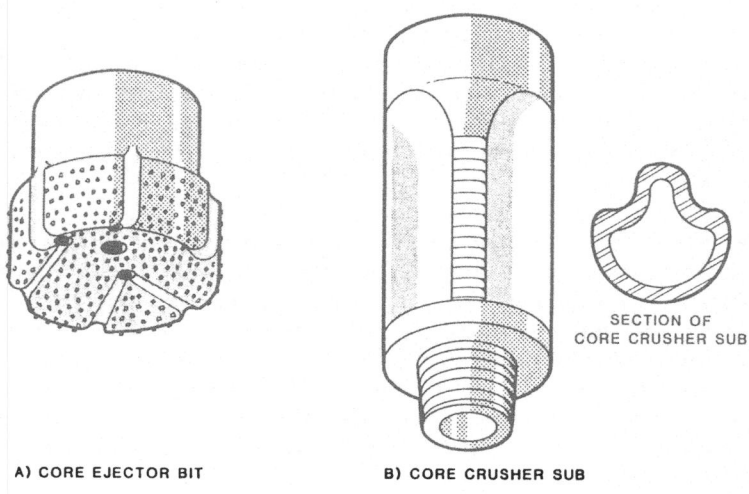

SECTION OF
CORE CRUSHER SUB

A) CORE EJECTOR BIT   B) CORE CRUSHER SUB

Figure 2-7. Core Ejector Bit and Crusher Sub

2.16 SIDEWALL CORING

Sidewall cores taken from the borehole wall after drilling generally produce samples of inferior quality to bottomhole cores. They are smaller in size and contain material which has been close to the exposed borehole wall for some time prior to the core being taken.

However, sidewall cores can commonly be obtained in less rig time and at less cost than bottomhole coring. Sidewall core points may also be selected after the hole is drilled and logged, reducing the risk of unnecessary coring of inappropriate zones.

2.17 WIRELINE CORE GUN

The sidewall coring gun, or Chronological Sample Taker (CST), is lowered into the hole on a logging cable and a sample of the formation is taken at the desired depth. This is done by shooting a hollow "bullet" into and pulling it out of the wall of the hole (Figure 2-8).

There can be twelve, twenty-four or thirty bullets per gun. By combining guns, up to seventy-two cores can be obtained during one run. If an electric log has been obtained previously, a spontaneous potential (SP) or gamma ray (GR) curve run in conjection with the samples can position the samples by direct log correlation.

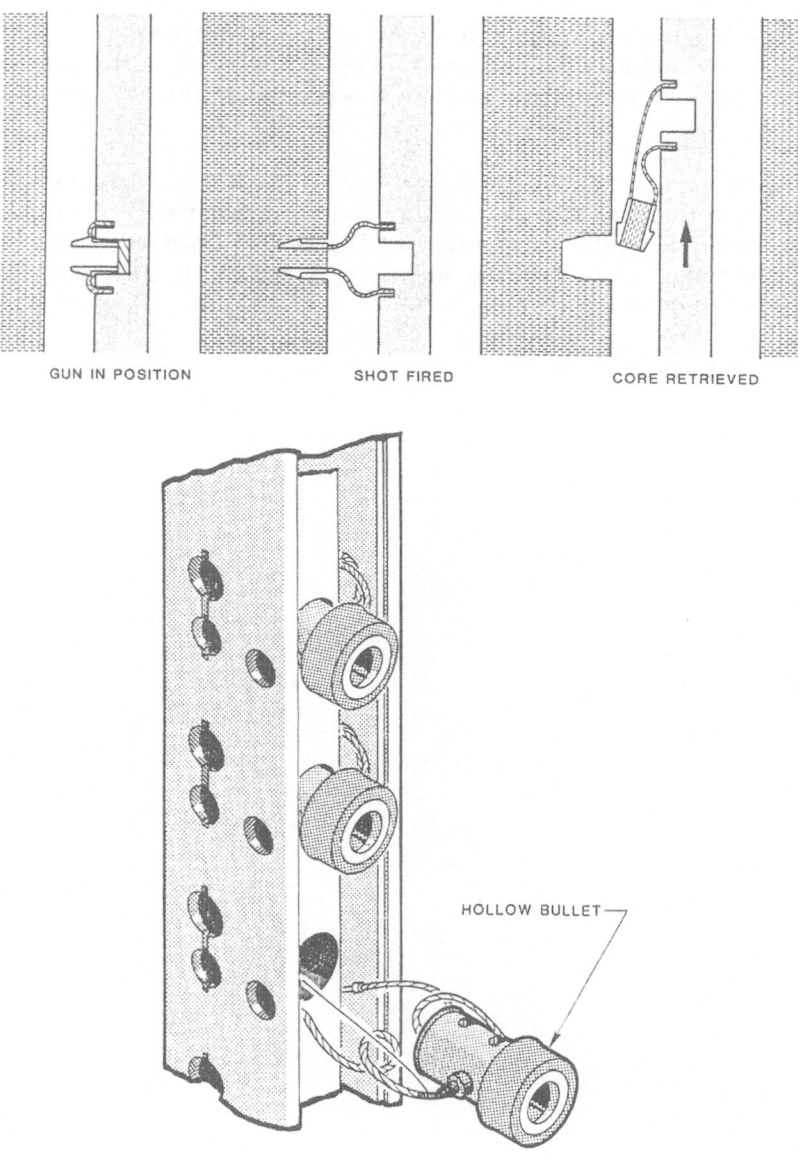

Figure 2-8. Sidewall Core (CST) Gun and Recovery Sequence

48

Sidewall cores taken with CSTs are small (1 x 2-1/2 inches), and in some cases the recovered material consists largely of mud cake. Sidewall coring is usually unsuccessful in very hard rocks. Nevertheless, cores of this type provide a means of examining the rock in portions of the section on which information may be extremely scanty. Sidewall cores are sometimes taken with the intention of evaluating the porosity, permeability and saturation characteristics of the rock. However, because compaction or fracturing occur as the bullet enters, the results are inevitably less reliable than those from a conventional core.

In some cases the logging geologist may be requested to perform gas analysis on sidewall cores. The CST gas analysis procedure is identical to that performed on gas samples from DSTs and wireline formation test tools except for the method of collecting the sample.

The escape of gases when CSTs are transported to the surface and when removed from the core chambers influences the gas reading somewhat, but with proper allowances CST gas analysis can provide useful information. The test is conducted after the cores have been removed from the core chambers and placed into small glass containers. The procedure is described in detail in Section 3.

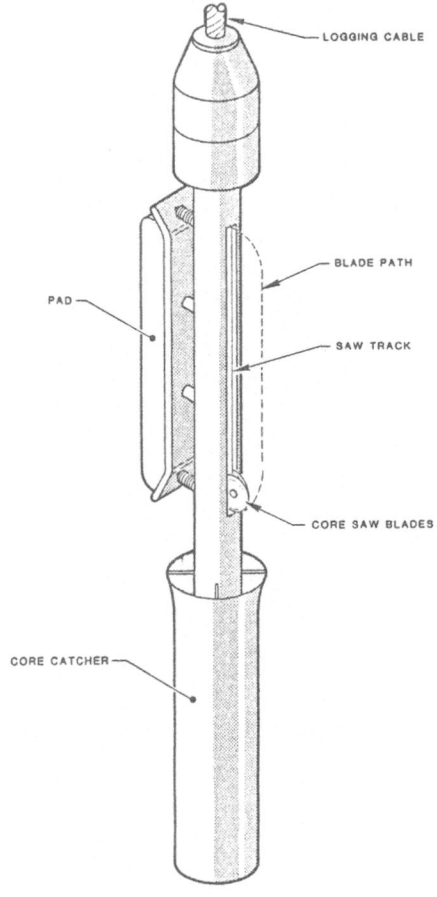

Figure 2-9. Wireline Core Slicer

2.18 WIRELINE CORE SLICER

In well-consolidated formations, it is possible to obtain larger sidewall samples using a diamond core slicer or Tricore tool run on a logging cable (see Figure 2-9). The tool is run to the appropriate depth for coring and a rubber pad is activated, forcing the tool against the borehole wall. Two diamond-edged circular saws mounted at 60° to each other move up the tool and out into the formation, cutting a triangular section core. The core is up to three feet long and approximately 1-1/2-inches on each side, with tapered ends following the path of the blades out of and into the tool.

On completion of the cut, the pad is retracted and the core falls into a core catcher below. The core catcher is divided into sections, allowing up to four cores to be caught on a single run in the hole.

Core slices are of unsuitable shape and size for core analysis but can yield valuable geological data. Unfortunately, unless the formation is extremely well-consolidated and ideal borehole conditions exist, the core slicer cannot be used. Irregularity or rugosity of the borehole wall will prevent contact, resulting in little or no blade penetration and only a fragmentary core. Poorly consolidated or friable formation will not survive the vibration during coring and the fall into the core catcher, and only unconsolidated fragments will be recovered.

2.19 ROTARY SIDEWALL CORER

This is an unusual variation of the retrievable core barrel. The inner barrel contains two universal joints, allowing it to be deflected by an internal whipstock out of the side of the outer barrel and into the formation at an angle of 20° from vertical.

Unlike a bottomhole core barrel, the outer barrel does not have a core bit attached. It is attached to the drillstring via a swivel, allowing it to remain stationary while the string rotates. The inner barrel has a rotary cutter head at its lower end and a splined joint at its upper end, allowing it to rotate with the drillstring.

Coring is accomplished by hanging the drillstring in the slips with the outer barrel positioned at the depth of interest. Rotation and circulation are commenced with mud hydraulics providing the force on bit required to feed the inner barrel out of the outer barrel and into the formation. When the barrel is full, circulation ports are exposed and a drop in pump pressure is seen. The inner barrel is recovered on a wireline, the outer barrel moved to a new core depth and a new inner barrel lowered into place.

Cores of this type are only one inch in diameter and up to one foot in length. They are of only marginally better quality than cores from a CST and require greater rig time and cost to obtain. For this reason, rotary sidewall coring is not regularly performed.

3

RETRIEVAL AND SAMPLING OF CORES

3.1   GENERAL

In the evaluation and analysis of cores, there is usually a division of responsibilities at or away from the wellsite. On some occasions the logging crew may be required to recover, evaluate the core, sample it and perform complete core analysis. In such cases it is recommended that two logging geologists work together, dividing the various tasks between them in order to complete the job speedily and efficiently. Since coring operations involve relatively short on-bottom intervals separated by longer-than-normal trip times, it is possible to do this without logging geologists working for excessively long periods or an unfair distribution of work. The Trailer Captain should organize duty and rest periods so that extra manpower is available when required. If, in his opinion, extra personnel or overtime credits are justified, the client's representative and local office should be informed.

At the opposite extreme, a coring crew may be assigned to the wellsite by the client or a service company. They may perform core analysis at the wellsite or process and sample the core for later laboratory analysis. In these circumstances, the logging geologist is required to be a geological observer only, and will have plenty of time available to fully evaluate the core.

The most commonly encountered situation falls between these two. Although not required to perform core analysis, the logging geologist's responsibilities involve the retrieval and packing of the core and the selection and sealing of samples for later core analysis. Depending upon the length of core recovered and the type of sampling required, the workload may require only one logging geologist or both working together.

This section of the manual covers this latter situation. Section 4 of the manual covers core analysis procedures. If core analysis is to be performed, the duties described in these two sections should be divided between the logging geologists in such a way as to allow coordination and maximum efficiency. If an oil company geologist is present at the wellsite, he should be included in the logging crew's planning so that he plays a coordinated part in the effort, avoiding duplication and misunderstandings.

3.2   CONVENTIONAL CORES

3.3   LONG-TERM PLANNING

When coring is anticipated on a well, the planning and ordering of supplies should begin as early as possible. Where long supply lines exist, it is desirable that necessary material, and a reasonable excess, be at the wellsite or at a nearby supply base, at spud. Figure 3-1 shows the types of material that must be available. The coring and sampling program for the well should indicate which of these will be necessary and what material is already available at the wellsite or locally.

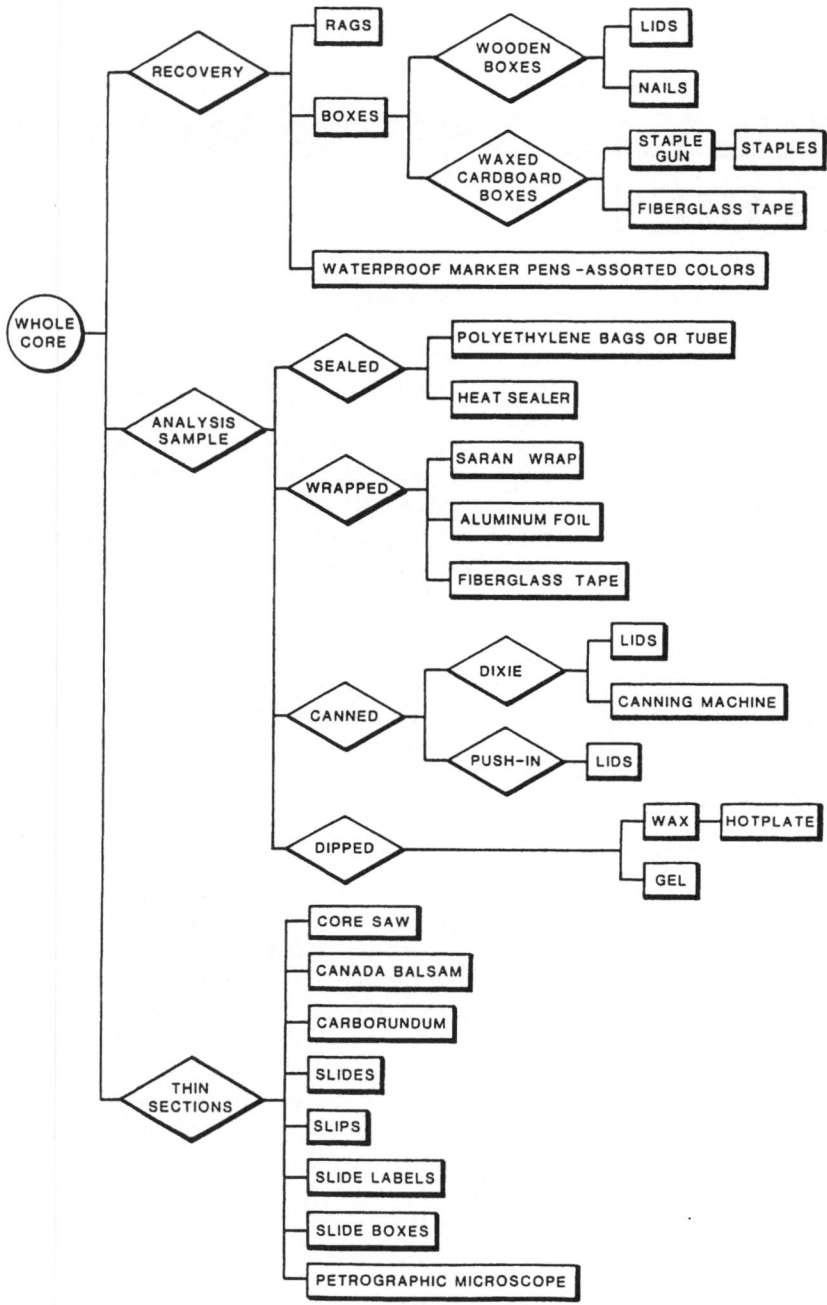

Figure 3-1. Supplies Required for Coring Operations

If any requirement is not well-defined, discuss it with the oil company representatives and obtain definite instructions. If necessary, assist them in ordering locally-unavailable supplies. Some of the required items are in the standard unit inventory, but not in sufficient quantity to handle extensive coring. If so, order extra supplies with an explanation of the special requirement.

If the rig floor crews are unfamiliar with coring operations, it may be worthwhile to brief them beforehand of the importance of careful handling of the core. With the permission of the client and the rig toolpusher, the Trailer Captain should attend a rig safety meeting, explain why cores are to be cut, how they will be recovered and analyzed and what part each crew member is to play in the operation. This is a good time to explain why the core should not be washed with the water hose. Some inexperienced roughnecks will do this, thinking they are being helpful!

As supplies arrive at the wellsite they should be inventoried and stored in a safe, dry place. Remember that, offshore, wood is a rare commodity and wooden core boxes will need to be guarded. Onshore, Saran Wrap and Aluminum Foil sometimes disappear on rig barbecue night. Note any shortages in supplies and send out reminders or reorders.

3.4 PICKING A CORE POINT

The most common reason for bottomhole coring is to determine the reservoir character- istics of a known potential reservoir. Picking a core point is therefore a matter of normal stratigraphic correlation. The approximate depth of the top of the formation will be known. When approaching that depth, correlation with offset logs is attempted to pick a point as close as possible above the formation top.

The most useful tools in picking tops in this way are unsmoothed drilling exponents, such as nx and dxc. These exponents have lithology and porosity dependent characters, minor but distinctive variations when plotted, allowing good correlation of overlayed plots from different wells. When plotted to a suitable scale, they also correlate well with offset sonic logs (see Figure 3-2).

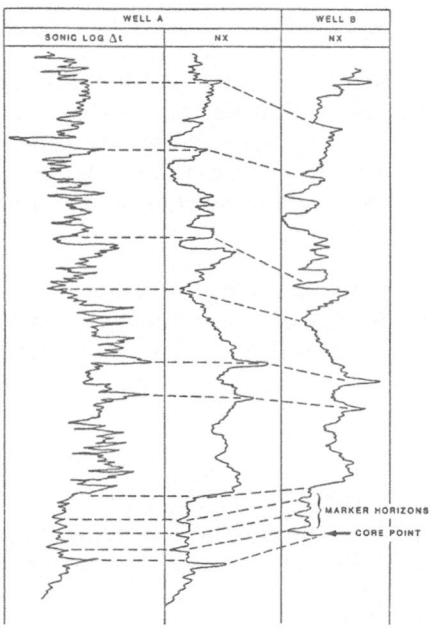

Figure 3-2. Correlation of Drilling Exponents with a Sonic Log

Comparison of the sonic and nx logs from Well A indicates which aspects of log character are genuine formation effects and may be looked for as marker horizons on the nx log of Well B. Data variations which do not occur in both the sonic and the nx log are probably erroneous, or system fluctuations, and cannot be expected to occur in Well B.

If drilling exponent logs are not available, correlation between rate of penetration and sonic log is possible. Remember, when comparing rate of penetration with a sonic log that variations due to changes in drilling parameters should be recognized and discounted. This is even more important when attempting to correlate rates of penetration directly from two different wells.

Often, drilling exponent or rate-of-penetration correlation will be inconclusive and lithological, or gas data will be required to confirm a core point. This may require interruptions to drilling in order to circulate returns. This is expensive in terms of rig time and should be minimized. However, the alternatives -- lost data due to drilling beyond the core point, or commencing coring too high -- may be even more expensive.

Coring is less commonly performed on wildcat wells, but if it is planned, the criteria for selecting core points will be included in the logging instructions. If not, the Trailer Captain should determine them from the client's geologists. Typical instructions may be as follows:

1. Below 8100 feet, circulate out all drilling breaks after cutting five feet.
2. If break contains oil shows, P.O.H. to core.
3. If no shows, drill ahead.
4. If break continues to a total of 15 ft, circulate returns and P.O.H. to core.

3.5 SET-UP TO CORE

Although only a limited selection of core bits and barrels will be available at the wellsite, the drilling supervisor will, nevertheless, require geological information in deciding the set-up of the coring assembly. Factors such as the type of core bit to use, optimum drilling parameters, positioning of the core catcher for best recovery, the length of the core to be attempted, bottomhole assembly design, and required mud properties must all be considered.

The logging geologist can assist by supplying information about the hardness and abrasiveness of the formation, the degree of consolidation, the likelihood of fractures (both natural and induced) and the possibility of hole problems due to swelling clays or geopressure.

3.6 LOGGING THE CORE

Coring provides the best available geological and lithological information when and if the core is recovered. While coring, the conventional formation logging is sparse and of poor quality. While some of this data (for example lithology) can be reconstructed from measurements and observation of recovered core, some (for example ditch gas) cannot. No data is available from sections of the core which are lost from the barrel. When coring is under way, take special care in logging to obtain all available information to supplement that gained from the core or to replace that which may be lost with the core.

3.7   Depth and Footage

It is common, on the trip-out prior to coring, to strap-out of the hole and obtain a reliable Total Measured Depth for the core point. Thereafter, if all core barrels are filled and have one hundred percent recovery of whole core (all broken ends fitting together conclusively), the depth and footage of all subsequent cores can be reliably and consistently known. Unfortunately, these ideal conditions rarely coincide and, as a result, depth discrepancies commonly occur during coring operations. Causes of depth discrepancies are discussed in Paragraphs 3.8 through 3.10.

3.8   Low Weight-on-Bit used in coring results in a greater surface hookload and therefore more stretch in the drill pipe held in tension. If Measured Depth is determined from the pipe tally, without consideration of pipe stretch, a depth error of as much as two to three feet will result. For example, a core barrel is run to bottom and low weight-on-bit applied. Due to pipe stretch, the Kelly is two feet higher than the driller expects from his pipe tally. The driller may report the first two feet of penetration as reaming to bottom. In fact, he is already coring. If the core barrel obtains full recovery, the error will be found. If it does not, the error may continue and accumulate on future core runs.

3.9   Gradual Changes in Weight-on-Bit will cause equally gradual change in pipe stretch and kelly height. A decrease in kelly height, due to increasing weight-on-bit, may be falsely interpreted as penetration being made without kelly movement. These errors can occur when starting and ending a core and when connections are made.

These small errors may result in inaccurate estimates of footage made by the core bit. For example, a 30-ft-long core barrel may appear to have cut 32 ft of core. Alternatively, it may appear to have jammed with 28 ft in the barrel, when it is in fact full.

Again, if one hundred percent recovery of a full core barrel is made, the error may be corrected. If not, the error can accumulate.

3.10   Lost Core from the bottom of the barrel, when making connections or when pulling out of the hole, may subsequently be drilled up, or reenter the core barrel as whole core or as broken fragments. The loss or gain of sections of core may be indicated on recovery by the presence of adjacent pieces of core whose ends cannot be made to fit together. The actual length of core lost or gained cannot be accurately determined, and so length of core recovery is not a reliable estimate of length of core cut.

Commonly, the depth and lengths of cores finally reported result from compromises between the driller's figures, the logging geologist's figures and the recovered core length. The figures are accepted as being imprecise. The logging geologist should take the figures used by the oil company drilling supervisor on drilling reports as the correct figure for use on all logs, reports and labeling of cores.

When a depth and footage for the core have been determined, the top and bottom of the core are noted on the formation evaluation log wth horizontal lines crossing the core column and penetrating the depth column by two millimeters, that is, almost, but not quite to the center of the depth column (see Figure 3-3). After the core is recovered and measured, the recovered, upper portion of the core interval is inked-in (see Figure 3-4).

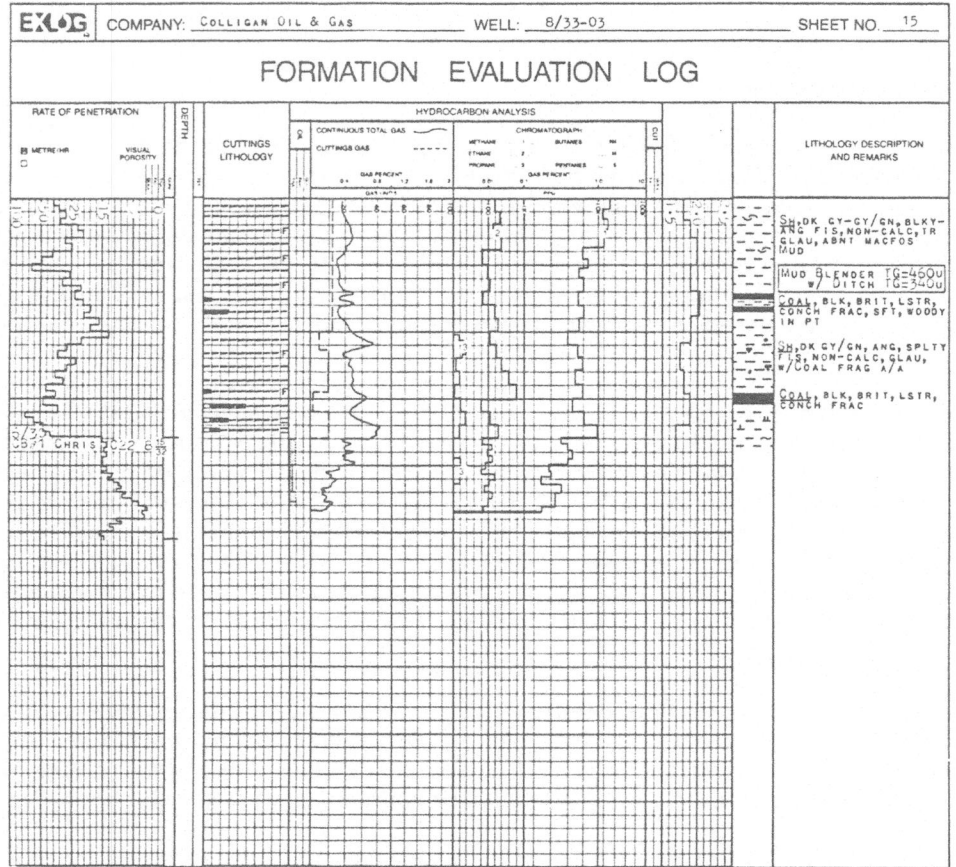

Figure 3-3. Core Log Prior to Recovery

3.11 Rate of Penetration

The imprecise in-depth measurement while coring will be reflected by equivalent inaccuracies in rate of penetration over the interval. Rates of penetration will be erroneously high when weight-on-bit is being applied, and low when weight-on-bit is being drilled off. However, for the remainder of the cored interval, when weight-on-bit is held constant, good rate-of-penetration data may be recorded (although minor depth errors may be included).

With relatively constant drilling parameters, minor variations in rate of penetration can be diagnostic of hardness variations in the rock, planes of weakness and fractures. Rate of penetration should be plotted in the smallest possible depth increments permitted by the vertical scale of the log (Figure 3-3). At the most commonly used vertical scales, one foot or half-meter increments are possible.

3.12 Gas Evaluation

As discussed in Section 1, the formation cut by the core bit will be extensively flushed and will contain limited residual hydrocarbons. The reduced volume of formation cut by the annular core bit at lower than normal rates of penetration will further reduce the volume of gas liberated to the mudstream.

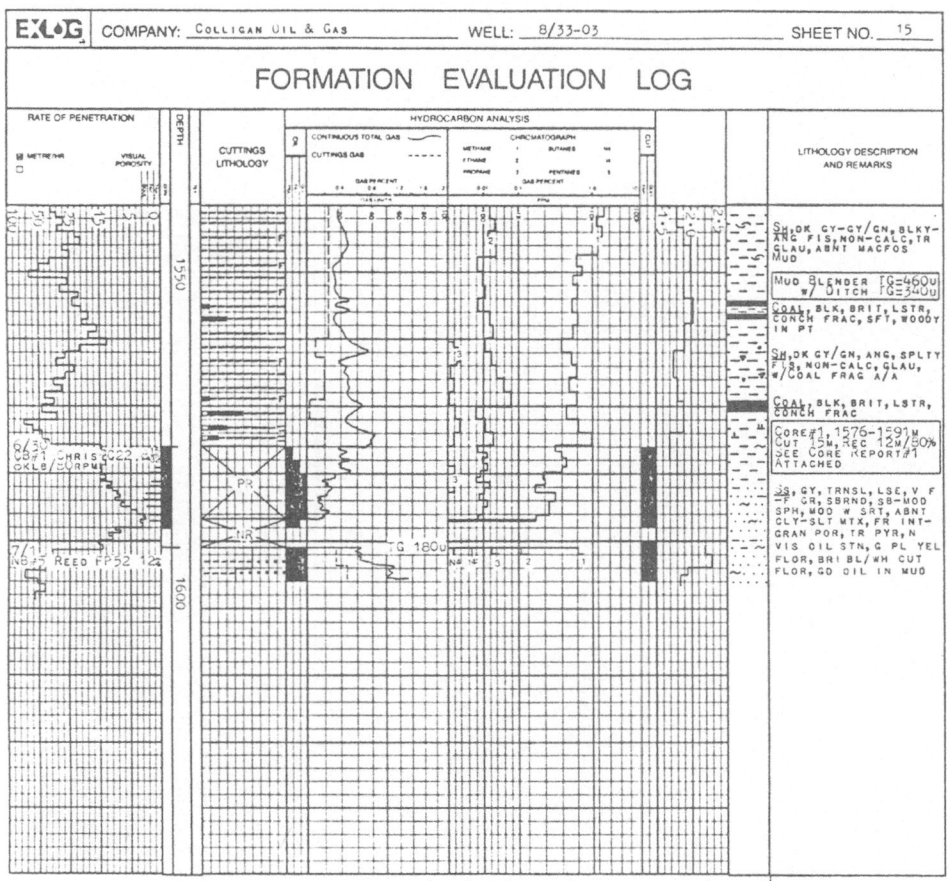

Figure 3-4. Core Log Completed After Recovery

Conversely, the lower-than-normal mud flowrates used when coring will increase the volume of gas per volume of mud arriving at surface. Sufficient gas will be contained in the drilling mud at surface for the logging of background gas and shows to be worthwhile, although not directly comparable with shows from the previously drilled section (see Figure 3-3).

When logging a cored interval, remember that the low flowrate through the ditch will require adjustment of the gas trap position and elevation. Lower the level of the gas trap and move it closer, if possible, to the flowline to obtain maximum flow-through and extraction efficiency.

Continuous total gas and chromatograph data are logged as normal. Lower rates of penetration should allow greater depth resolution in these. Low rates of cuttings recovery and pulverization by the diamond core bit will prevent cuttings gas analyses from having any meaning, and this curve should be projected to zero, using a solid line. Remember that a solid line to zero indicates that no analysis was made, as opposed to no gas in the sample.

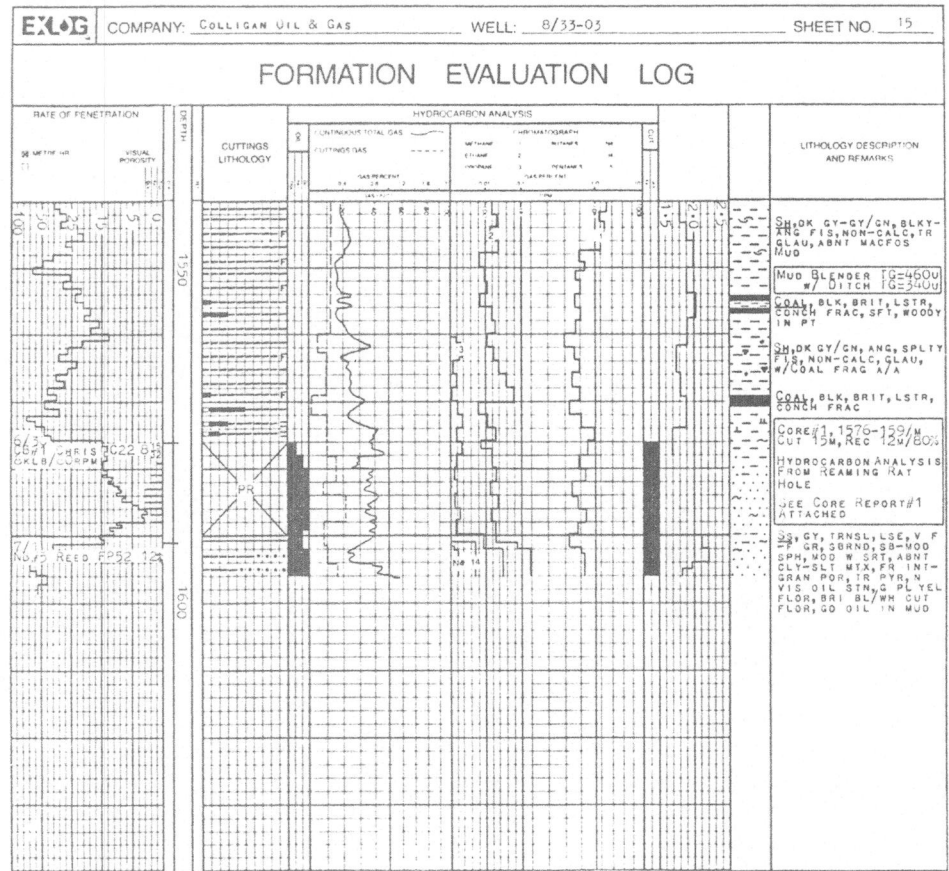

Figure 3-5. Oil and Gas Relogged from Rat-Hole

Circulation after coring is kept to a minimum to prevent washing and loss of the core. Thus, data from the lower portion of the core cannot be logged before tripping-out. Because of the low mud flowrate and hence long lag time, this may represent several feet of core. Because the perturbation of the mud column in tripping the large diameter core barrel and the low liberated gas volume relative to contamination and gas produced from the borehole wall, this gas data cannot be reliably logged after tripping back into the hole (LAT). The data is therefore lost. The ditch gas and chromatograph plots are projected to zero at the depth of last returns prior to tripping-out the core barrel.

If the core bit used is undersize, better gas analyses may be available from ditch and cuttings, after coring, when the rat-hole is reamed. If the oil company geologist permits, the gas analyses for the cored interval may be logged in pencil. When the rat-hole is reamed, the pencil plots can be erased and new ditch gas, cuttings gas and chromatograph logs plotted in ink. In this case, a note should be made in the Remarks column that the gas analyses were logged when reaming (see Figure 3-5).

3.13 Oil Evaluation

Flushing and pulverization of cuttings combine to prevent reliable oil evaluation during coring. However, because of the possibility of loss of the core, fluorescence tests on cuttings and mud samples must be performed while coring. The results (natural color, color and intensity of fluorescence, and cut if any) should be recorded on the worksheet for later reference. An oil evaluation may be lightly penciled on the log (see Figure 3-3). Later, oil evaluation may be inked in, based upon these tests and from recovered core (see Figure 3-4). Gaps in the oil evaluation, due to loss of circulation samples and core, may be filled with data logged when reaming the rat-hole (see Figure 3-5).

3.14 Cuttings Lithology

Cuttings returns while coring will be of low quality and quantity. Nevertheless, samples must be caught and bagged to maintain the completeness of trade sets, and as a safeguard against lost core. A brief sample description may be noted on the worksheet for future reference.

While coring, no lithological notations should be made on the log. After the core is recovered, the Cutting Lithology column may be annotated as 'Poor Returns' (PR) or 'No Returns' (NR), as applicable. On the basis of cuttings, core, reaming and other data, the Interpreted Lithology column may be drafted (see Figure 3-4 and 3-5). The Visual Porosity Bar Graph may also be added to the Rate of Penetration column after the inspection of recovered core.

3.15 MUD TESTS

The drilling mud circulation system should be stabilized prior to coring. Additions and treatments of the mud active system during coring should be limited to those necessary to maintain stable properties. If this is done, tests can be performed which can provide estimates of filtrate invasion of the core. Such estimates at various points in the core may assist in evaluating core permeability and pore water salinity.

3.16 Salinity

The procedure for determining the salinity of mud filtrate is described in Appendix B of The Field Geologist's Training Guide (Exlog, 1985). Salinity titrations should be performed from the suction line every thirty minutes, and an average value established. If there is wide variation about the average, individual values must be down-lagged to the appropriate depths in the core. Otherwise, the average value may be used.

If the formation pore water salinity is known from resistivity log analysis of formation testing, estimates of filtrate invasion at various points in the core may be made

$$C_c = C_m(M) + C_w(1 - M) \qquad (3\text{-}1)$$

$$M = \frac{(C_w - C_c)}{(C_w - C_m)} \qquad (3\text{-}2$$

where

C_c = salinity of core water, ppm

C_m = salinity of mud filtrate, ppm

C_w = salinity of formation pore water, ppm

M = proportion of mud filtrate in core water, unitless (0-1.0)

$$I = M * S_w \qquad (3\text{-}3)$$

where

I = proportion of mud filtrate in core porosity, unitless (0-1.0)
S_w = core analysis water saturation, unitless (0-1.0)

Note that "I" is, determined here as a surface measurement and does not account for water expulsion from the core during recovery to surface.

For example:

mud salinity, C_m = 80,000 ppm

formation water salinity, C_w = 110,000 ppm

core water salinity, C_c = 98,000 ppm

water saturation, S_w = 60%

$$M = \frac{(110,000 - 98,000)}{(110,000 - 80,000)} \qquad (3\text{-}2)$$

= 0.4 or 40%

I = 0.4 * 0.6 $\qquad (3\text{-}3)$

= 0.24 or 24%

3.17 Nitrate Ion

If a nitrate ion test kit is supplied in the logging unit it can provide a reliable estimate of filtrate invasion even when formation water salinity is unknown.

The principle of the method depends upon the almost unknown occurrence of nitrates in formation waters in rocks of sufficient age to be petroliferous. Thus, if the mud filtrate is 'spiked' with a nitrate salt in a known concentration, the measured concentration of nitrate in water recovered from a core or formation test will be a measure of the proportion of mud filtrate, of known concentration, mixed with formation water, of zero concentration. Analogous to equation (3-2):

$$M = \frac{(N_w - N_c)}{(N_w - N_m)} \qquad (3-4)$$

where

N_c =   nitrate ion concentration of core water, ppm

N_m =   nitrate ion concentration of mud filtrate, ppm

N_w =   nitrate ion concentration of formation pore water, ppm

but

N_w =   0

Therefore:

$$M = \frac{N_c}{N_m} \qquad (3-5)$$

And solving equation (3-1) for C_w, Formation Pore Water Salinity

$$C_w = \frac{\left[C_c - C_m (M) \right]}{(1 - M)} \qquad (3-6)$$

For example

mud salinity, C_m =   80,000 ppm

core water salinity, C_c =   98,000 ppm

mud nitrate ion concentration, N_m =   270 ppm

core water nitrate ion concentration, N_c =   85 ppm

water saturation, S_w =   60 %

$$M = \frac{85}{270} \tag{3-5}$$

$$= 0.375 \text{ or } \underline{37.5\%}$$

$$I = 0.375 * 0.60 \tag{3-3}$$

$$= 0.19 \text{ or } \underline{19\%}$$

$$C_w = \frac{[98,000 - (80,000 * 0.375)]}{(1 - 0.375)} \tag{3-6}$$

$$= \underline{108,800 \text{ ppm}}$$

3.18 PREPARATON FOR CORE RETRIEVAL

As soon as the kelly is broken-off and the trip-out begun, the logging geologist must begin preparation for retrieving the core. A catching set of core boxes must be prepared, the necessary tools and materials assembled and a work area cleared and ready for handling the core.

3.19 Work Area

Even when core analysis is not being performed, the logging unit is not a convenient place to handle and view long sections of core. A work area should be found adjacent to the logging unit where the whole core can be laid out, preferably at bench height, not on the floor, with sufficient space for two or more people to work around it. Ideally, the work area should be under cover and well-lit (cores can be pulled at night and in the rain) and be in a 'Safe-Area', allowing electrical devices, such as the U.V. Light Box and Core Drill to be set up and used.

Available in the work area should be:

- Sufficient clean core boxes for the full core barrel, plus a few spare
- Labeled sample bags and envelopes
- A supply of clean rags
- Sufficient wrapping, sealing or canning supplies
- A measuring tape
- A supply of marker pens
- Spare work gloves
- Worksheet or note pads
- A copy of the core sampling and shipping instructions

3.20 Rig Floor

When the drill collars reach surface, the logging geologist should begin to assemble the required material on the rig floor for core retrieval. These are

- A catching set of core boxes
- A hammer
- A broom
- Work gloves
- A supply of clean rags
- A note pad and pencil on a clipboard

The catching set of core boxes are required for transport of the core from the rig floor to the work area where the core will be cleaned and transferred to new boxes. The catching set may then be cleaned out and reused.

The catching set consists of sufficient boxes to contain the full core barrel plus about forty percent excess since the core will not come out of the barrel in convenient lengths and should not be broken unnecessarily. The boxes should be clearly labeled with Top, Bottom and Number. This will avoid errors on the possibly poorly lit rig floor (see Figure 3-6).

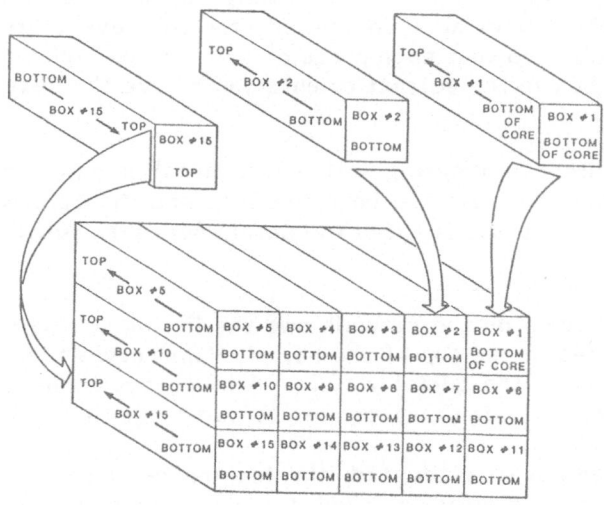

Figure 3-6. Labeling and Stacking of Catching Set

The catching set should be stacked in appropriate order in an area of the rig floor, out of the way of the floor crew and close to the area where core will be retrieved. A clear area in the vicinity should be allocated for the stacking of filled core boxes. On an offshore rig, the work area may be so far from the rig floor that a crane will be used to move the core boxes. In this case, a pallet should be placed on the rig floor upon which to stack the filled core boxes.

If the boxes are to be carried singly to the work area, finalize at this time who will be doing that job. Remember that a 60-ft core will require approximately twenty-five catching boxes which, when full, will weigh over 40 lb each. This is too large a job for the logging geologist to perform alone and would take him too much time when he should be evaluating and sampling the core. Although assignment of tasks in core retrieval should have been performed in crew briefings (see Paragraph 3.3), this is a good time to finalize with the driller all details of who will be assigned to carry core boxes to the work area. Ideally, the boxes should be moved as they become filled, keeping the core recovery area clear and allowing the oil company geologist or the other logging geologist to begin evaluation at the work area.

Finally, call the oil company geologist, other logging geologists or anyone else who will be required when the core is removed.

3.21 REMOVING CORE FROM THE BARREL

When the core barrel reaches surface, the outer barrel is set in the slips. The ball is retrieved from the check valve and a lifting sub made up on the inner barrel. The inner barrel is then lifted from out of the outer barrel with the elevators and moved over to the core retrieval area.

The core catcher is removed and, with the core held in place with a support pin, core tongs are attached (see Figure 3-7a). The core tongs allow the core to be gripped so that it can be slid out of the barrel at a controlled rate to prevent breakage. Rate of retrieval should be controlled so that each piece of core can be picked up, briefly inspected and placed in the core box in the correct orientation before the next piece is allowed out of the barrel.

If the core is jammed, light hammering with a soft metal hammer on the barrel may free it. Excessive hammering may damage both the core and the core barrel. If hammering cannot free the core, or if it is suspected to contain hydrogen sulfide, it must be pumped out.

The inner barrel is laid down on the catwalk and the lifting sub removed. A rubber plug, or core pusher, is inserted in the top of the barrel. This serves to prevent contamination by separating the core from the pumping fluid. A pump-out connector is made up on the inner barrel and high pressure air or water used to pump-out the core (see Figure 3-7b).

Whichever method of removal is adopted, the logging geologist's activities during core retrieval remain the same. In either method, the rate at which core is removed from the barrel should be governed by the rate at which the logging geologist can retrieve and box the core (see Paragraph 3.22). Do not allow yourself to be hurried at the risk of core pieces being inverted or boxed out of order. In addition, two basic rules must always be followed:

1. DO NOT PLACE ANY PART OF YOUR BODY BELOW A HANGING CORE BARREL OR IN FRONT OF A CORE BARREL THAT IS BEING PUMPED OUT. THIS APPLIES EVEN WHEN THE BARREL IS THOUGHT TO CONTAIN NO MORE CORE!

2. ALL PERSONNEL INVOLVED SHOULD UNDERSTAND THE FUNCTION OF EACH OTHER AND REMAIN WITHIN CLEAR VIEW OF EACH OTHER AT ALL TIMES. THE DRILLER, THE PERSON OPERATING THE CORE TONGS OR PUMP-OUT PRESSURE VALVE, AND THE LOGGING GEOLOGIST MUST BE POSITIONED SO AS NOT TO BLOCK EACH OTHER'S VIEW OF THE CORE BARREL. ALL OTHER PERSONNEL SHOULD STAND WELL CLEAR.

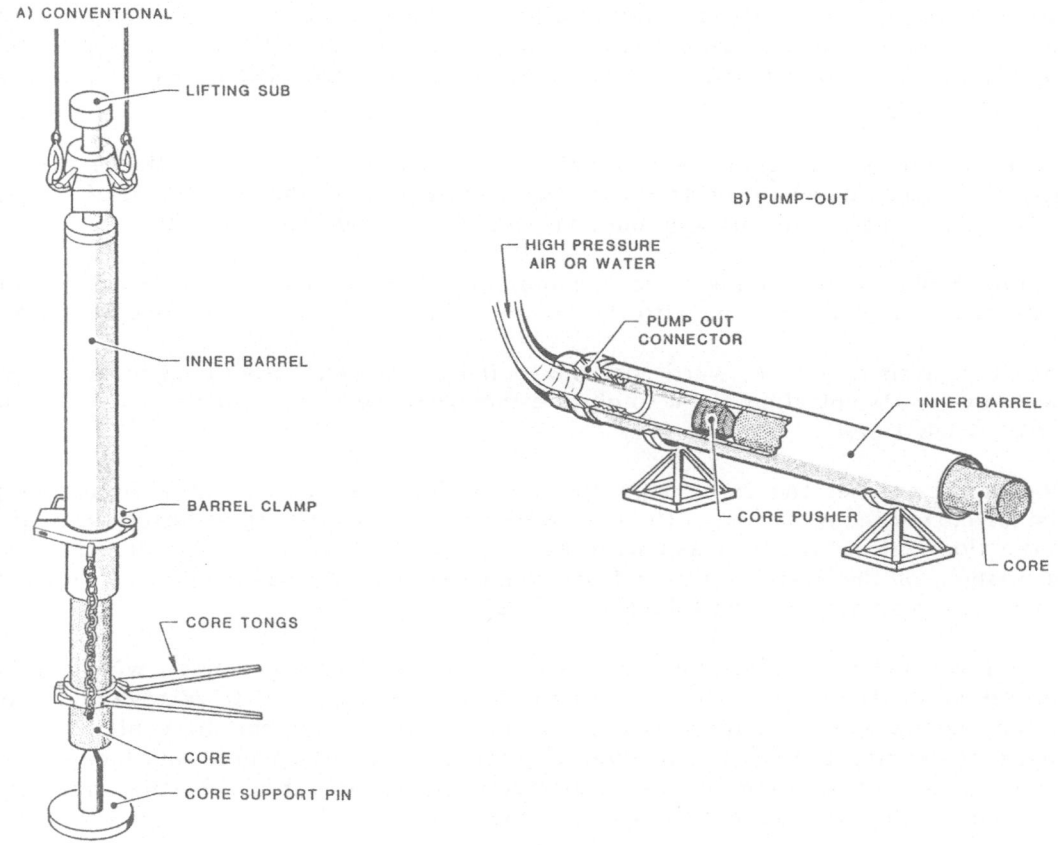

Figure 3-7. Removing Core from the Barrel

3.22 RETRIEVING THE CORE

As the core is removed from the barrel, draw each piece away from the barrel using the broom. Have the floor hand operating the core tongs, or pressure valve hold back the rest of the core until you are ready to go on.

Holding the core carefully, to avoid crumbling unconsolidated material or accidentally inverting it, wipe the core free of drilling fluid with a clean rag. Give the core a brief visual examination. Notice any petroleum or other odors, bleeding of oil from fractures or intergranular porosity. Look for 'blowing' gas. This may be seen and heard as small bubbles popping on the circumference or broken faces of the core, or if minor may be felt as a "prickling" sensation under the fingers when the core is held. Another indication of fluid bleeding or blowing from the core is when a piece of core remains wet even after it has been wiped dry several times. Make brief notes on the clipboard. For example

"BOX #3 -- NEAR TOP, OIL BLDG FROM HORIZ FRACS"

Place the core piece in the correct box in the correct orientation and go on to the next.

The core will usually be broken naturally into lengths convenient for boxing. Avoid using the hammer to break the core unless absolutely necessary. Do not worry if long pieces overhang the end of a box slightly, or if small pieces do not quite fill a box. This can be adjusted later.

Core pieces should be boxed in the order they come out of the barrel. If the core is vertically fractured, one-half of the core may come out of the barrel ahead of the other. They should be boxed this way and adjustment to correct fit made later.

Do not allow broken core fragments to accumulate under the barrel. As they fall from the barrel they should be swept up with the broom and put into the appropriate core box.

Personnel delegated to pass forward core boxes and carry away filled ones must always stand behind you. Do not allow them to block either your view or the driller's view of the core barrel at any time.

When the rabbit reaches the core tongs, the core is fully retrieved. Have the catching set of boxes moved immediately to the core work area. Remove surplus boxes out of the way of the floor crew. Again, make stern warnings to the crew about washing the core or even washing of the area of the rig floor while the recovered core remains exposed. Also, warn crew members against taking souvenirs!

During retrieval and removal to the work area, the oil company geologist, who may be more interested in stratigraphy than in physical properties may want to remove pieces of core to the logging unit for closer inspection. Try to discourage this by explaining the importance of keeping the core in recovered order until it is measured and labeled. If this is not successful, ask him to take and return pieces one at a time and mark the missing piece by stuffing rags into the space in the box.

3.23 BOXING THE CORE

Processing the core in the work area is best performed by two geologists working sequentially as follows (see Figure 3-8):

- Geologist One -- recovers core and oversees removal to work area
- Geologist Two -- reboxes and labels the core
- Geologist One -- follows him, removing and processing core analysis samples
- Geologist Two -- follows him, sampling and examining for geological evaluation
- Geologist One -- follows him, closing the boxes and preparing the core for shipping

Given the normally slow rate of coring and tripping of core barrels, two logging geologists can work together this way without the need for excessively long work periods.

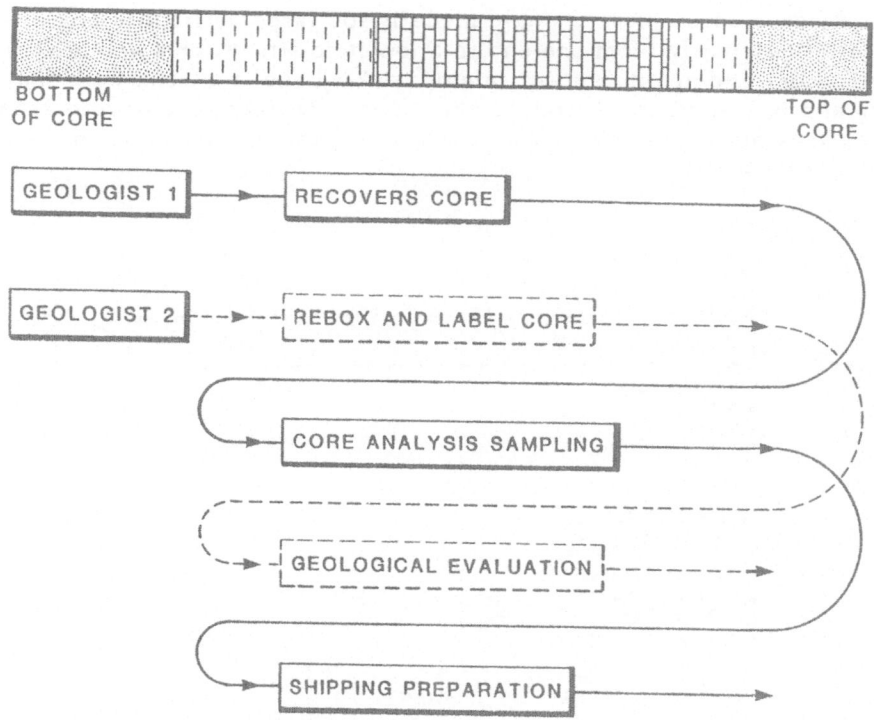

Figure 3-8. Sequential Core Processing by Two Geologists

At the work area, the first requirement is to transfer the core to clean core boxes, establish the fit between pieces and measure the core.

Most oil companies require core boxes to be numbered in the order of recovery from the barrel. Thus, box number one will contain the bottom of the core and subsequent numbers will contain upper parts. The core will be transferred to the clean core boxes in the same order that it was put into the catching set (see Figure 3-9a). This convention will be followed in the remainder of this manual.

Some European oil companies prefer the core boxes to be numbered in the reverse order, that is, in the order that the core was cut. Thus, box number one will contain the top of the core. The correct numbering sequence should be established before reboxing begins (see Figure 3-9b).

Boxes should be clearly labeled 'Top', 'Bottom' and box number. Box number one should be labeled 'Bottom of Core' or 'Top of Core', as applicable. Further labeling is not necessary until the core is measured.

Each piece of core should be taken from the catching box, wiped with a rag and obvious features noted on a worksheet pad, for example, oil bleeding, gas blow or visible macro structures. Then place the core piece into the new box and proceed with the next piece. If a piece of core is too large to fit in the space available in a box, start a new box. Do not break a core unless a single piece is too large to fit in a box alone. In that case, break off sufficiently from one end to allow it to fit. Rubble and broken fragments should be wiped dry and put in a clean sample sack. Label the sack with the well name, core number and box number. Estimate the number of feet of core represented by the rubble in each sack and write this on the label. Place the sample sacks in the appropriate place in the core box.

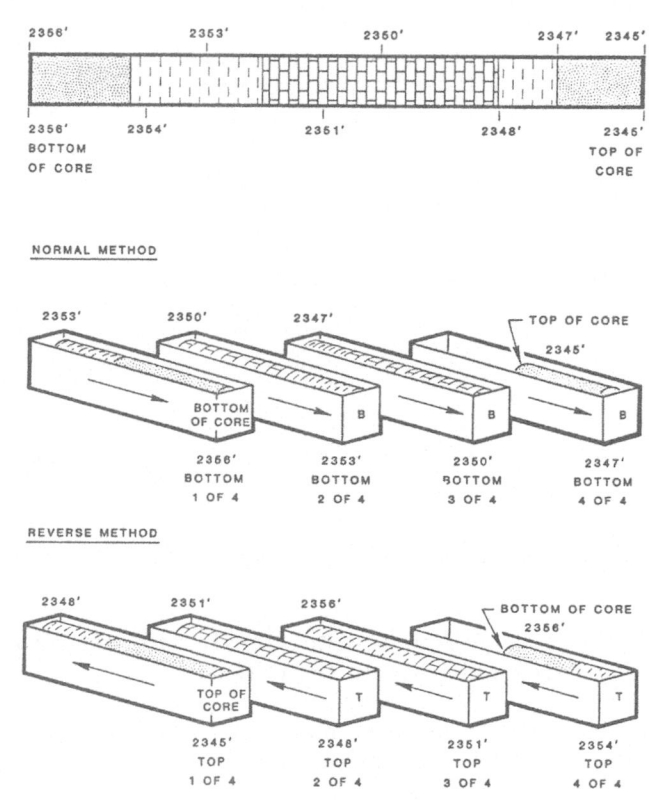

Figure 3-9. Core Box Numbering

When the box is almost full (three to six inches short of being full), rotate the core piece and push them together to get the best possible fit. Take a black and a red marker pen and mark two lines down the length of the core with the red pen to the left (see Figure 3-10). This ensures that core pieces can in future always be oriented.

If the core is fractured vertically, mark black and red lines on both halves of the core. Examine the quality of fit between adjacent pieces of the core. If there is no fit, mark the two ends with double chevrons. If there is some similarity of shape but not an exact fit, mark the two ends with single chevrons. If there is good fit, make no marks. This preserves a record of core continuity if pieces are later removed or damaged in transit. If there is no fit or poor fit, briefly examine the ends of the core to ascertain whether this is due to solution or deposition on a natural fracture, or to some mechanical effect such as breakage, washing or loss of the core ends. Make suitable notations on the worksheet.

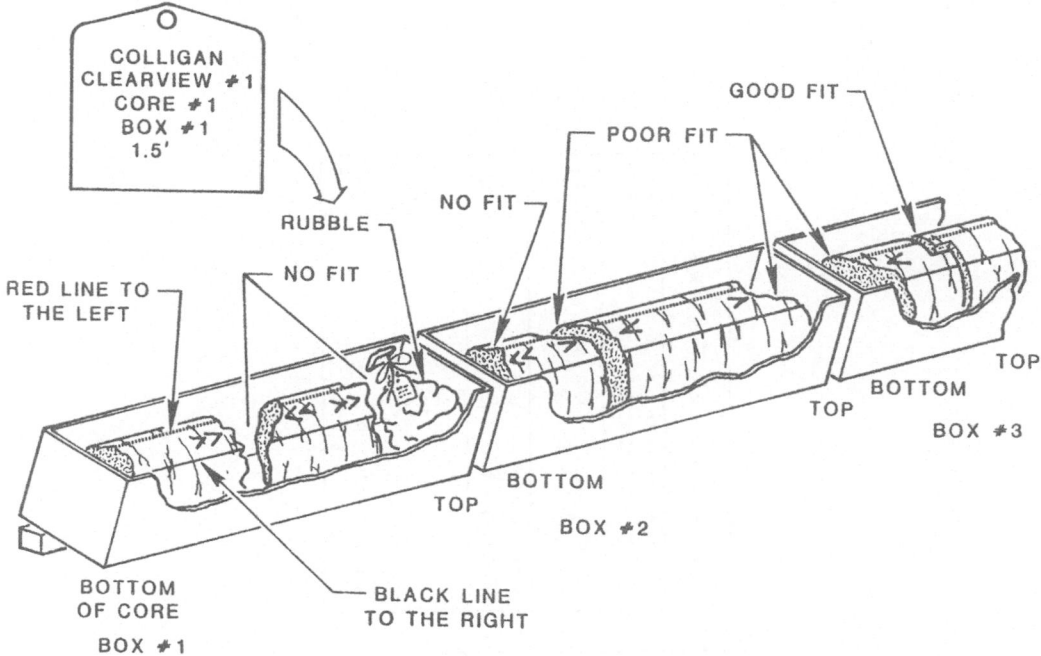

Figure 3-10. Labeling of Core Orientation, Fit and Rubble

NOTE

Conventions for marking core orientation and fit may vary between oil companies. The above convention is recommended in the American Association of Petroleum Geologists 'Sample Examination Manual'.

With the ends of the core pieces pushed together, measure the core length in the box. Keep a tally of these (see Figure 3-11). Move the pieces of core apart slightly and stuff rags at the top and bottom of the box and between each piece of core. This prevents damage to structure on the ends of the core pieces. If the core shows signs of bleeding gas, take one small piece from the box, seal it in a 'zip-lock' plastic bag and return it to the box for later gas analysis (see Paragraph 3.30). The sample should not occupy more than 1/20 of the volume of the bag; otherwise, expanding gas will overinflate and burst the bag. Now proceed to the next box.

When all of the core has been reboxed and measured, the core recovery and the interval in each box can be calculated. Remember that reboxing and measuring started at the bottom of the core, but the normal assumption is that any core lost is lost from the bottom. It is therefore impossible to determine the depth represented by the bottom of the <u>recovered</u> core until the total length of core is known. Figure 3-11 is an example worksheet for tallying and calculating core recovery and intervals.

Core #6
Cut, from: 6621·8 ft
     to   : 6650·7 ft

Box #	MEASURED LENGTH			Box INTERVAL	
	CORE	EST. RUBBLE	TOTAL	FROM	TO
16					
15					
14					
13					
12					
11					
10					
9	1·84	2·0	3·84	6621·8	6625·6
8	2·58	—	2·58		6628·2
7	2·66	—	2·66		6630·9
6	2·88	—	2·88		6633·8
5	2·02	1·0	3·02		6636·8
4	2·57	—	2·57		6639·4
3	2·45	1·0	3·45		6642·8
2	2·21	0·5	2·71		6645·5
1	2·68	—	2·68		6648·2

	CUT	RECOVERED
BOTTOM	6650·7	6648·2
TOP	6621·8	6621·8
FOOTAGE	28·9	26·4
% RECOVERY		91%

Figure 3-11. Core Recovery Tally Sheet

3.24 SAMPLING FOR CORE ANALYSIS

For core analysis it is necessary to select short pieces, four to six inches long of whole core from which core plugs will be taken.

The usefulness of data obtained from core analysis is limited by the care that goes into the selection and preservation of samples. Therefore, proper sampling of a conventional core is one of the most important phases of the core analysis procedure. A sufficient number of samples must be taken so that data representative of the complete zone can be obtained. The samples must be selected immediately after removal from the core barrel, and they must be adequately guarded by proper preservation against alteration of the samples or of the contained fluids. Where insufficient sampling occurs in conjunction with a hasty lithologic description, then the net productive thickness, transition zones, and fluid contacts cannot be definitely defined -- possibly resulting in erroneous interpretations.

Field sampling of core should represent the best possible practices because the value of core analysis is limited by this initial operation. The objective of a standard field core sampling procedure is twofold:

- To obtain samples which will give results representative of the formation

- To obtain samples through uniform procedures so that the results will be independent of the sampler

When the core has been retrieved, two criteria must be dealt with:

- The selection of representative samples from each core

- The wrapping and preserving of the core samples quickly enough to prevent loss of fluids from within the core or the absorption of foreign fluids by the core

3.25 Sample Selection

The selection of samples is fairly simple for relatively uniform formations. However, where a formation contains widely varying lithology and heterogeneous porosity types (such as conglomerates, weathered cherts, vugular or fractured carbonates, and interlaminated shales and sands), proper selection of representative samples requires greater care. The logging geologist should follow a fixed sampling procedure at the well location.

Sampling frequently varies with the particular job, but the following general guides are listed.

- Ordinarily, in fairly thick homogeneous sands, one sample per foot is taken. It is sometimes sufficient to take one sample every two feet. The interval will be specified in the logging instructions.

- The sampling intervals should be reduced if the reservoir section is nonuniform and varies greatly over short distances in properties such as porosity, permeability and oil content. It may be advisable to take the samples closer than one per foot to represent the core. Remember, however, that even adjacent samples may vary drastically in their properties.

- In thick permeable zones with variable lithologic properties, obtain at least three samples -- one each from the top, middle and bottom portions.

- In horizons where alternate thin shale and sand members occur, sample every sand member over two inches in thickness.

A reasonable approach is to select samples from the center of representative sections of the core based on visual examination. For example, in one case a sample may be considered as representative of only three inches of core, whereas in another case a sample may be representative of a section one foot or more in length.

When selecting samples, refer to the descriptive notes made in recovery and reboxing the core, and where possible, add to them. Sections of the core from which bleeding or blowing of fluids occured may represent good porosity and permeability. However, remember that zones with the best permeability will have been extensively flushed by drilling fuid and gas expansion and may be totally depleted and inert by the time they reach surface.

Although obtaining representative samples is the prime consideration, try to minimize breakage of the core. Wherever possible, take already broken pieces of core of approximately the correct size. This is convenient when performing wellsite core analysis, since a large piece of core can be removed from the box, plugs taken, and it can be returned. When selecting samples for shipping to a core laboratory, a compromise must be struck between obtaining sufficient representative analysis samples and retaining sufficient core for geological evaluation.

3.26 Sample Preservation

Technique required to preserve core samples for testing depends on the length of time for storage or transit and the nature of the tests desired.

- Double wrapping: A suitable method for short-term preservation is to double- or even triple-wrap the samples in aluminum foil or Saran Wrap.

- Canning: This is not considered the best method because an air space remains in the can, making it possible for pore space fluids to evaporate into it.

- Wrapping and canning: This method is much better than just canning. As much air space as possible should be eliminated by using abundant nonabsorbent material.

- Freezing in dry ice: Samples preserved by this method can be kept for a long period of time without the fluid saturations or other properties of the core being affected, though facilities for this method are rarely present at the wellsite.

- Wrapping and paraffin wax coating: This is a very good method, and if the wax coating is properly applied and protected against breakage, it will protect the core sample for an extended time.

- Plastic coating (Core Dip Gel): This method is better than coating with wax because the plastic coating is much more durable than the wax coating.

The method recommended by Exlog is a triple wrapping procedure using Saran Wrap, Aluminum Foil and polyethylene sleeve. This method combines one of the best modes of preservation with one of the quickest procedures. Thus, the maximum number of samples may be sealed in the minimum time, ensuring optimum and relatively uniform sample quality.

A selected sample of appropriate size is first double-wrapped with Saran Wrap. It is then wrapped again with Aluminum Foil. If core analysis is to be performed at the wellsite, the wrapped sample should be labeled with the footage interval and may then be returned to its appropriate place in the core box. See Section 4 for the next stages in the procedure.

If the samples are to be shipped to a core laboratory, further wrapping is required. The sample is strapped once around the circumference and once longitudinally with fiberglass tape. It is then marked with red and black lines to indicate orientation and labeled with:

- Oil company name
- Well name
- Core number
- Sample number
- Sample depth interval

The whole sample is then heat-sealed in a polyethylene tube.

If the whole core is being shipped to a nearby laboratory, the sealed sample can be returned to the correct location in the box.

If the sealed samples are to be shipped separately, they should be packed carefully in a wooden or metal box with straw, rags or newspaper separating the samples. A shipping inventory should be prepared. One copy should be enclosed with the samples, one given to the oil company representative and one retained in the logging unit until receipt is acknowledged by the laboratory. The inventory should contain:

- Name of the laboratory and responsible personnel
- Oil company, well name and location
- Core number
- Total number of samples
- List of sample numbers and intervals
- Type of analysis required
- Mud type and water loss
- Mud salinity and nitrate ion concentration
- Name and address to whom reports are to be sent

The spaces in the core boxes from which samples have been removed should be stuffed with rags and 'Core Analysis Sample', and the length removed written on the inside of the box at that point.

3.27 GEOLOGICAL EVALUATION

3.28 Macroscopic Examination

Following the brief examinations made and noted during the recovery and boxing of the core and when selecting core analysis samples, a complete macroscopic examination should now be made. This should include:

- The lithology and thickness of major lithological units
- The nature and dip of lithological boundaries
- The size and dip of bedding, sedimentary and diagenetic structures
- Gradations within beds
- The spacing and dip of natural fractures and partings
- Surface condition of natural fracture surfaces

- Type, amount and distribution of secondary porosity
- Presence of hydrocarbon staining or odor

Make a rough sketch of the core showing lithological boundaries and major structures.

3.29 Microscopic Examination

Keeping core breakage to a minimum, take small chips of core at one foot intervals and at points of special interest. Place the chips in labeled sample envelopes.

Each core chip should be given a complete microscopic examination and described according to :

- Rock type
- Color
- Induration
- Mineralogy
- Grain size
- Grain shape
- Grain texture
- Sorting

- Cementation
- Matrix
- Microstructures
- Porosity type
- Porosity size
- Accessory mineral
- Fossils

Combining these descriptions with the macroscopic examination, a complete core description report can be prepared. This is drafted onto a sheet of clear vellum for later attachment to the end of the Formation Evaluation Log (see Figure 3-12). The core description should contain macroscopic and microscopic features. If the core has sufficient visible macrostructure, a drawing of the core may be added to supplement the description.

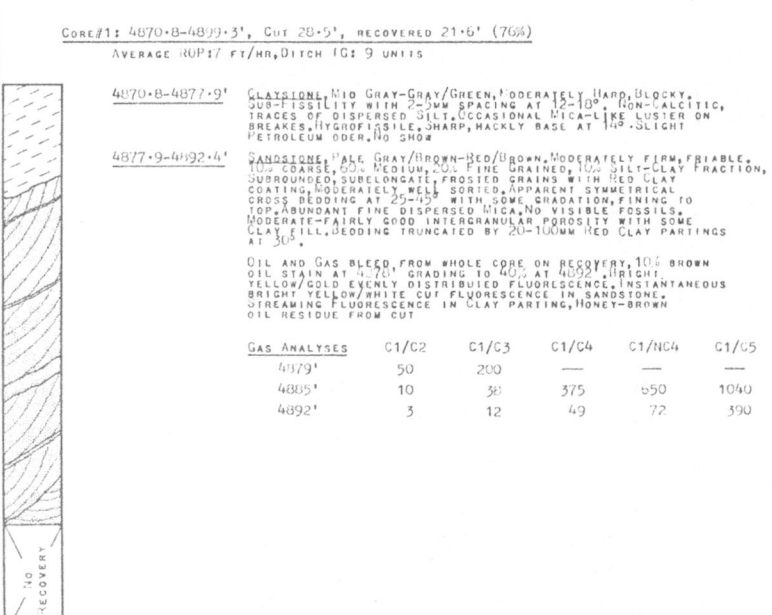

Figure 3-12. Core Description Log

The core chips are returned to their labeled envelopes and added to the washed and dried sample set.

3.30 Hydrocarbon Evaluation

View the complete core under ultraviolet light, and make notes of the location, distribution, color and intensity of fluorescence.

Individual core chips should each be subjected to complete oil evaluation tests, noting

- Petroleum odor
- Color and distribution of oil stain
- Color, intensity and distribution of fluorescence
- Type and rate of cut
- Color and intensity of cut fluorescence
- Color of cut residue

Add this information to the core description (see Figure 3-12).

If bleeding gas samples were previously sealed (see Paragraph 3.23) they may now be analyzed in the chromatograph.

Take a sample, note the depth and, using a hypodermic syringe, make a hole and withdraw 10 cc of gas. Remove the syringe and seal the hole with scotch tape.

Take a plastic squeeze bottle and squeeze it to reduce its volume by about a quarter. Inject the gas sample into the squeeze bottle while allowing the bottle to slowly expand to its full volume. This prevents escape of gas from the bottle during injection. Seal the top of the squeeze bottle with a cap or scotch tape.

Set up the chromatograph for a single sample, hold (BACKFLUSH disabled) and manually inject the sample from the squeeze bottle in the same manner as a manual calibration (see the appropriate Technical Manual for the type of chromatograph to be used). If off-scale readings occur, adjust the ATTENUATOR settings and repeat the test.

Return the sample to the core box, flush out the syringe and squeeze bottle with clean air and proceed with the next sample.

If you do not have a hypodermic syringe, the sample can be taken directly into the squeeze bottle. Make a small hole in the plastic bag. Compress the squeeze bottle to reduce its volume by 25%, push its spout into the hole in the bag and allow it to expand to full volume. Remove the squeeze bottle and seal the hole with scotch tape. Compress the squeeze bottle to half its volume and allow it to expand to full volume. Repeat this two more times. This should give a suitable dilution for analysis.

If you do not have a squeeze bottle, a plastic wash bottle or diswashing liquid bottle can be substituted.

Regardless of the sampling method used, this analysis cannot be directly related to the quantity of gas in place. However, it does give a useful estimate of the proportion of gases present. For this reason, the results should be reported on the Core Report as ratios, C1/C2, C1/C3, C1/IC4, C1/NC4, etc. If the unit cailbration gas does not contain pentane, a reasonable estimate can be obtained by using the normal butane calibration factor for a catalytic chromatograph or four-fifths (0.8x) of the calibration factor for an F.I.D. chromatograph. Results should be added to the core description (see Figure 3-12).

3.31 SPECIAL SAMPLES

Occasionally it may be necessary to prepare special geological samples for reflected or transmitted light examination. These are polished slabs and thin sections respectively.

This is not a standard operation and the equipment and procedures vary. The following is an outline for general information. Specific instructions will be given to crews required to perform the operations.

3.32 Core Slabbing

The core slabber is a table-mounted, electrically-driven, circular saw using diamond-tipped saw blades (see Figure 3-13). A number of models are available. When operating them be sure to follow the manufacturers instruction manual, especially with regard to safety. Wear safety glasses and use all machine shields as directed.

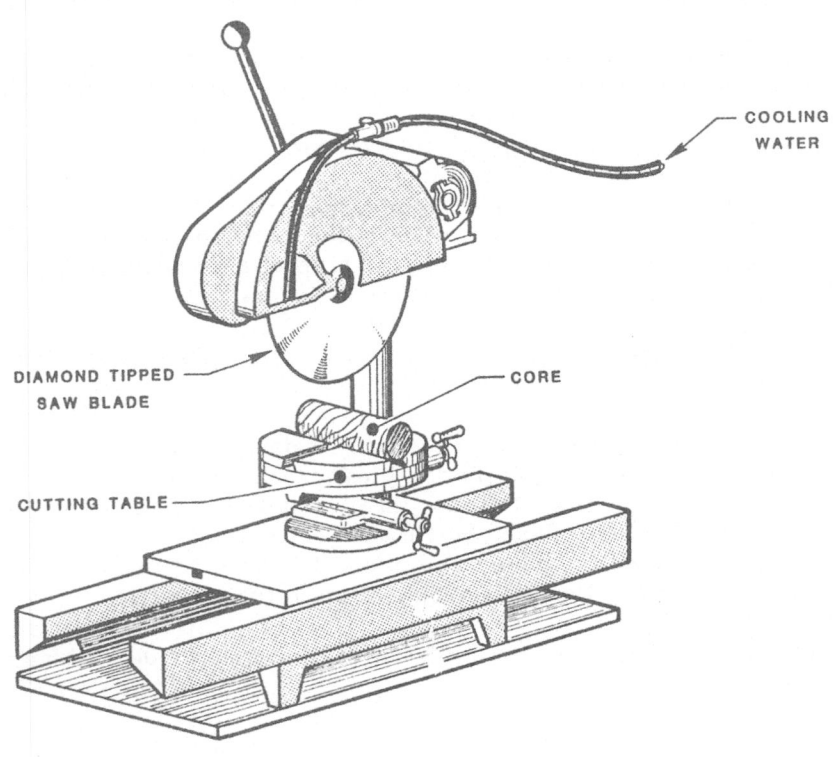

Figure 3-13. Core Slabber

Using the core slabber, the core may be cut horizontally or vertically into flat-faced slabs suitable for examination under the microscope (see Figure 3-14). Visual examination may be enhanced by polishing the slab faces using carborundum powder on a glass lapping plate (see Paragraph 3.33). After polishing, the surface may be etched by immersion in very weak hydrochloric acid (1%; one part 10% acid to nine parts distilled water) for ten to fifteen seconds.

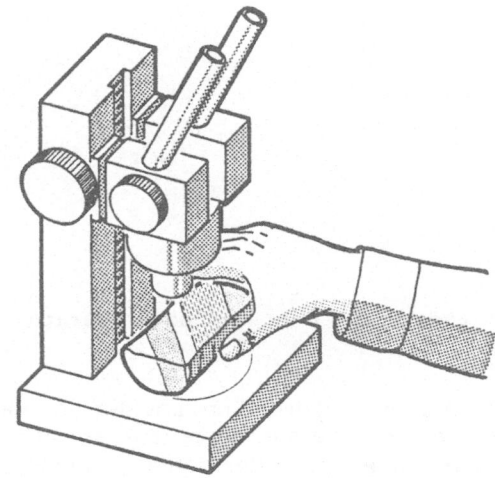

Figure 3-14. Visual Examination of Etched Slab

3.33 Thin Sections

Thin sections for examination with transmitted light using a petrological microscope can be prepared from thin core slabs, core chips or cuttings. The sample is mounted onto a glass slide, ground and polished to the standard petrological thickness of thirty microns (0.03mm), although thickness is not so critical as when a polarizing stage is to be used.

Heat Canada Balsam, the standard medium, in a pan on the hot plate until it melts to a viscous fluid. Dip a glass rod into the Canada Balsam, pick up a small amount and transfer it to a glass slide. Spread the liquid to form a thin film and embed the rock sample securely in it. Proceed to the next sample, leaving the slides for at least thirty minutes to cool and harden.

In grinding and polishing the slide, perform every operation gently and carefully! Even slight shocks may result in the slide or sample being broken or the sample detached.

Attach the slide to a glass or metal rod using a small suction cup. If this is not available, a small spot of 'super glue' can be used. Use only enough so that the rod may later be broken-off without breaking or marking the slide (see Figure 3-15).

Mount the rod in a retort stand and lower it until it is almost touching the grinding wheel surface. Start the grinding wheel at a low speed and lower the slide until contact is just made. Increase the speed of the wheel and grind the sample until contact is lost. Slow down the wheel and lower the sample slightly. Repeat this process until a flat sample approximately forty to fifty microns thick is obtained.

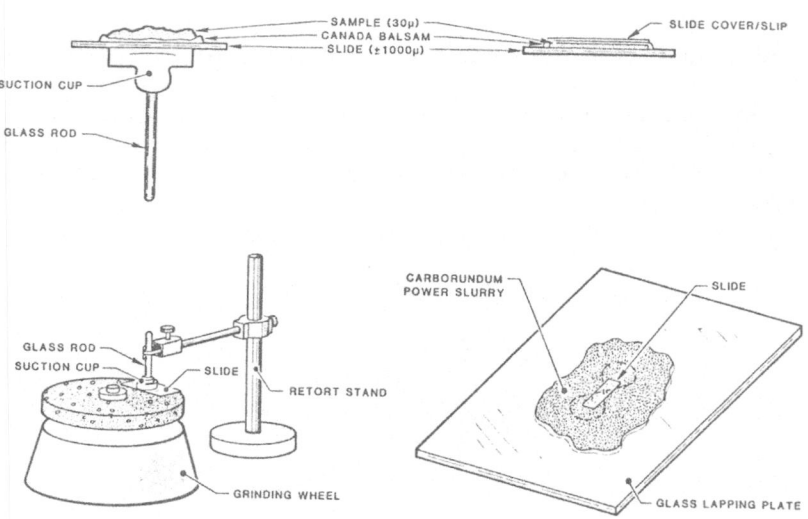

Figure 3-15. Thin Section Preparation

Prepare a thick slurry of coarse carborundum powder in water on the glass lapping plate. Apply light pressure on the slide and move it in a figure-of-eight pattern over the plate to ensure uniform, level polishing. Periodically check the slide thickness with a micrometer screw gauge. Be careful not to overtighten the micrometer, which can very easily crack the slide.

As the sample is ground, change to finer grades of carborundum powder so that a final thickness of thirty microns (total thickness minus slide thickness) is achieved with the finest grade of carborundum (1000 grade) and a smooth polished sample surface. Drip a small quantity of Canada Balsam onto the sample, spread it to a thin film and attach the glass slide cover or slip. Attach a paper label to the slide with

- Well name
- Slide number
- Depth
- Orientation, if known
- Lithology

3.34 Acetate Peels

An alternative to thin sections, though not quite so useful, is the preparation of acetate peels. These can be produced easily and in large sections up to six inches or more. Two methods are available: (1) using a solution of cellulose acetate, or (2) the more practical method using a cellulose acetate film. The liquid form is made up as follows:

Parlodion	112	g
Butyl Acetate	1000	cc
Amyl Alcohol	40	cc
Xylene	40	cc
Ethyl Ether	12	cc
Castor Oil	12	cc

As an alternative to this rather offensive mixture, acetate films are available from 0.002 to 0.020 inches in thickness. These are softened with acetone and applied directly. Softening time varies with film thickness, and experimentation may be necessary.

Slab, polish and etch the core as described above, and then dry and leave it to cool. If the liquid acetate is used, pour it onto the rock surface and allow it to set for 24 hours. Alternatively, pour acetone onto the etched surface and roll a piece of acetate film, matte side (if one exists) down across the acetone-wet area. Drive the excess acetone away along the advancing film-rock contact, maintaining a slight "bead" of liquid along the contact by tilting the rock slightly toward the direction of film advance. The film should not be pressed down, and at least one inch of margin of film should be left untouched by acetone around the softened area. Another method is to bow the film into a "U" shape, touch the center of the surface with the base of the "U", and gradually flatten from the center outward.

After setting (15 minutes for the film), gently peel the acetate from the rock surface, soak it in dilute hydrochloric acid for ten minutes or as necessary to remove adhering rock fragments, then rinse and dry. The resulting peel is mounted between glass sheets and may be used as a photographic negative and enlarged up to x50 (or even x80 in special cases) for grain size and textural studies.

3.35 SHIPPING THE CORE

After evaluation and sampling are complete, the core can be prepared for shipping.

Stuff more rags, as necessary, into the core boxes to prevent the core being damaged or disoriented in shipping. Along the edge of a wooden core box or on the inside of the top flap of a cardboard core box, mark depths at 1-ft intervals and mark the locations of core analysis samples that have been removed. Include in box number one, sealed in a plastic bag, a copy of the core analysis sample shipping inventory. Wooden boxes are then nailed shut, cardboard boxes are stapled and wrapped with fiberglass filament tape. If the boxes are to be exposed for some time, for example on the deck of a workboat, wrapping each box in plastic sheet and fiberglass filament tape is recommended.

The boxes are already labeled with box number "Top" and "Bottom". They must also now be labeled with an address and name of a responsible person to receive them. The boxes should also be labeled with

- Oil company name
- Well name
- Core number
- Depth interval in box
- Total number of boxes (Box #4 is augmented to read Box #4 of 8)

On a wooden core box this information should be on the side or the end of the box since the lid may later be removed and discarded. However, on a tight hole, the cored interval, or even the fact that coring is taking place, may be secret information. In this case, the information may be included inside the box.

A better alternative is to write the information in code. A simple substitution code is illustrated in Figure 3-16. A variation of this can be adopted for use in any 'tight-hole' situation. Using this particular version of the code:

Macy Exploration
Boulder Ranch No. 1
Core No. 3
9826.4 - 9880.7 feet

consisting of twenty-three boxes can be encoded

Code Number = 2 + 3 = <u>5</u>:   Use Code #5
                        Encode in <u>5</u> Digit Groups

CODE NUMBER — ENCODING

CLEAR	#1	#2	#3	#4	#5	#6	#7	#8	#9
A	T	Ø	J	H	Ø	/	E	H	S
B	A	N	T	P	6	Q	J	Q	B
C	U	A	3	X	G	F	O	I	L
D	B	1	A	5	T	1	T	R	T
E	V	Ø	K	A	7	7	Y	Z	C
F	C	B	U	I	A	L	3	4	I
G	W	2	4	Q	N	A	8	A	M
H	D	P	B	Y	1	W	A	J	U
I	X	C	L	6	H	R	F	S	D
J	E	3	V	B	U	G	K	B	N
K	Y	Q	5	J	8	2	P	K	V
L	F	D	C	R	B	8	U	T	Ø
M	2	4	M	Z	O	M	Z	C	1
N	G	R	W	7	2	B	4	L	E
O	Ø	E	6	C	I	X	9	U	J
P	H	5	D	K	V	S	/	Ø	O
Q	I	S	N	S	9	H	B	5	W
R	I	F	X	Ø	C	3	G	9	2
S	2	6	7	8	P	9	L	D	6
T	J	T	E	D	3	N	Q	M	F
U	3	G	O	L	J	C	V	V	P
V	K	7	Y	T	W	Y	Ø	1	X
W	4	U	8	1	-	T	5	6	3
X	L	H	F	9	D	I	-	-	7
Y	5	8	P	E	Q	4	C	*	9
Z	M	V	2	M	K	-	H	E	*
Ø	6	I	9	U	4	Ø	M	N	G
1	N	9	G	2	*	D	R	W	K
2	7	W	Q	-	E	Z	W	2	Q
3	O	J	Ø	F	R	U	1	7	Y
4	8	-	/	N	Y	J	6	/	4
5	P	X	H	V	L	5	D	F	8
6	9	K	R	3	5	*	I	O	-
7	Q	*	1	*	/	P	N	X	/
8	-	Y	-	G	F	E	S	3	H
9	R	L	I	O	S	Ø	X	8	R
-	*	1	S	W	Z	V	2	G	Z
.	S	M	2	4	M	K	7	P	5
	1	2	*	/		6	*	Y	A

Figure 3-16a. Substitution Codes: Encoding

Five Digit Code Groups =   MACY   Encoded Label =   OØGQM 7DVBI C03HI
                               EXPLO   2M6IJ BT7CM CØ2G1
                               RATIO   M2IZ4 MGIC7 M21ZE
                               N BOU   MF/*L LRSF//KZ5A
                               LDER   3MMMM
                               RANCH
                               NO. 1
                               CORE
                               NO. 3
                               9826
                               .4-98
                               80.7F
                               T

DECODING — DC NUMBER

	#1	CLEAR	#2	CLEAR	#3	CLEAR	#4	CLEAR	#5	CLEAR	#6	CLEAR	#7	CLEAR	#8	CLEAR	#9	CLEAR
	A	B	A	C	A	D	A	E	A	F	A	G	A	H	A	G	A	B
	B	D	B	F	B	H	B	J	B	L	B	N	B	Q	B	J	B	E
	C	F	C	I	C	L	C	O	C	R	C	U	C	Y	C	M	C	I
	D	J	D	L	D	P	D	T	D	X	D	1	D	5	D	S	D	N
	E	K	E	O	E	T	E	Y	E	3	E	8	E	A	E	Z	E	T
	F	L	F	R	F	X	F	3	F	7	F	C	F	I	F	S	F	Ø
	G	N	G	U	G	1	G	8	G	C	G	J	G	R	G	-	G	8
	H	P	H	X	H	5	H	A	H	I	H	Q	H	Z	H	A	H	F
	I	R	I	Ø	I	9	I	F	I	O	I	X	I	6	I	C	I	O
	J	T	J	3	J	A	J	K	J	U	J	4	J	B	J	H	J	1
	K	V	K	6	K	E	K	P	K	Ø	K	.	K	J	K	K	K	C
	L	X	L	9	L	I	L	U	L	6	L	F	L	S	L	N	L	G
	M	Z	M	.	M	M	M	Z	M	E	M	M	M	Ø	M	T	M	J
	N	1	N	A	N	Q	N	4	N	M	N	T	N	7	N	Ø	N	P
	O	3	O	E	O	U	O	9	O	S	O	Ø	O	C	O	6	O	U
	P	5	P	H	P	Y	P	B	P	Y	P	7	P	K	P	.	P	2
	Q	7	Q	K	Q	2	Q	G	Q	4	Q	B	Q	T	Q	B	Q	9
	R	9	R	N	R	6	R	L	R	-	R	I	R	1	R	D	R	A
	S	.	S	Q	S	-	S	Q	S	D	S	P	S	8	S	I	S	D
	T	A	T	T	T	B	T	V	T	J	T	W	T	D	T	L	T	H
	U	C	U	W	U	F	U	Ø	U	P	U	3	U	L	U	O	U	K
	V	E	V	Z	V	J	V	5	V	V	V	-	V	U	V	1	V	Q
	W	G	W	2	W	N	W	-	W	Z	W	H	W	2	W	7	W	V
	X	I	X	5	X	R	X	C	X	5	X	O	X	9	X	E	X	3
	Y	K	Y	8	Y	V	Y	H	Y	.	Y	V	Y	E	Y	P	Y	-
	Z	M	Z	A	Z	Z	Z	M	Z	A	Z	2	Z	M	Z	V	Z	L
	0	0	0	D	0	3	0	R	0	H	0	9	0	V	0	2	0	M
	1	Q	1	G	1	7	1	W	1	N	1	D	1	3	1	8	1	R
	2	S	2	J	2	.	2	1	2	T	2	K	2	-	2	F	2	W
	3	U	3	M	3	C	3	6	3	1	3	R	3	F	3	Q	3	4
	4	W	4	P	4	G	4	.	4	7	4	Y	4	N	4	W	4	.
	5	Y	5	S	5	K	5	P	5	B	5	5	5	W	5	3	5	S
	6	0	6	V	6	O	6	I	6	E	6	E	6	4	6	9	6	X
	7	2	7	Y	7	5	7	N	7	K	7	L	7	.	7	R	7	5
	8	4	8	1	8	W	8	S	8	Q	8	S	8	E	8	X	8	Y
	9	6	9	4	9	Ø	9	X	9	W	9	Z	9	O	9	Y	9	6
	-	8	-	7	-	8	-	2	-	2	-	6	-	X	-	X	-	Z
	*	-	*	-	*	4	*	7	*	8	*	A	*	P	*	Y	*	7
	1		1		1				1		1		1		1	4	1	

Figure 3-16b. Substitution Codes: Decoding

Notice that the information required by the decoder to select the appropriate code is openly available on the box and that by breaking the message into equal number groups with coded spaces, the original word lengths are obliterated. Using a different code number selector and by rescrambling the columns in Figure 3-16, other codes may be created as required.

A record of shipping and copies of all documents must be provided to the oil company representative at the wellsite (see Figure 3-17).

EXLOG

FORMATION LOGGING REPORT NO. ____

COMPANY SOUTHEAST OIL & GAS WELL MONTABELLO 1A

DATE 8/7/81 TIME 1800 HRS

DEPTH 2835 m LAST REPORT DEPTH 2817 m (0600 HRS)

REPORT BY J. KNOBLA (EXLOG) REPORT RECEIVED BY ____ SIGNED (OPERATOR)

BACKGROUND

Trip Gas: NONE @ ____

RIG OPERATIONS: CUT CORE #2, 2817.3 -2835.2, CUT 17.9 m, REC 12.8 m (72%)

Other Gas: NONE - CORING

LITHOLOGY: 4 m SHALE A/A, 8.8 m SST, F-MED GR, WELL SRT, W/OIL SHOW

Last Carbide: NO CARBIDE RUN w/ w: ____ v: ____

Lag @: -CORING spm: ____ min

Theoretical Lag: 97 min

ROP range: 2-8 m/HR avg.: 7

Mud Density: 0.12 S.G. Vis: 42 SEC

DITCH GAS range: Ø-Ø.2% avg.: Ø-15%

Salinity: 110,000 PPM Oil: —

CUTTINGS GAS range: — avg.: —

Logging Conditions, ~~Good/Fair~~/Poor: CORING

CHROMATOGRAPH range: C_1 Ø-0.2 ; C_2 TR ; C_3 TR ; IC_4 Ø ; NC_4 Ø

Estimated Over/Under Balance, ~~High/Mod.~~/Low: ____

SHOWS

Interval From/To	Lithology Visible Porosity	ROP	GAS SHOWS							Unwashed Free Oil, Live/Resid Inten. Color. Fluor	Washed % Stain Inten. Color. Fluor	Cut Inten. Color. Fluor	Oil in Mud	Salinity	Show Rating
			Ditch	Cutting	C_1	C_2	C_3	IC_4	NC_4						
2821.2 / 2830.1	SST, F-MED GR, WELL SRT, GD INTGR POR	8	0.2	–	0.2	TR	TR	–	–	TR OIL IN MUD	70% STN BRT YEL/ GLD FLOUR	INST BRT YEL/ WH	TR	110,000	GD
2830.1 / 2835.2	NO CORE RE														

Comments: CORE BEING PREPARED FOR SHIPPING.

CORE #1, 6 CORE BOXES AND 1 BOX CORE ANALYSIS SAMPLES SHIPPED AT 12:45 HRS ON BOARD M.V. MORCAMBE BAY. E.T.A. ABERDEEN 0900 HRS, 8/8/81. COPIES OF SHIPPING INVOICES ATTACHED.

CORE REPORTS #1, #2 AND FORMATION EVALUATION LOG SHEET #12 TELECOPIED TO SOUTHEAST, DALLAS, AT 1600 HRS.

THIS REPORT IS GOVERNED BY THE TERMS AND CONDITIONS AS SET FORTH ON THE REVERSE SIDE

Figure 3-17. Report of Core Processing and Shipping Details

3.36   **RETRIEVABLE BARREL CORES**

Although equipment differs, procedures for logging and processing these cores are similar to those used in conventional coring. Differences are discussed below.

3.37 LOGGING

Several cores may be obtained in a single run with retrievable core barrels. Although logged as a single, continuous bit run, each core should be logged as a separate core with footage and recovery reported accordingly.

3.38 SAMPLING

Recovery and sampling of retrievable barrel cores are identical to the procedures discussed for conventional cores. However, greater time restraints exist.

If only one inner barrel is available, recovery must be achieved as quickly as possible to allow it to be rerun. If two are available, the first core must be recovered, sampled and described in the time it takes to cut and retrieve the second.

3.39   **RUBBER SLEEVED CORES**

3.40 LOGGING

Rubber sleeved cores are cut in 2-ft increments with closure and drilling-off of the expansion joint and force-on-bit being applied with pump pressure (see Paragraph 2.12). Although this allows easy recognition of footage increments and depth, rates of penetration recorded over the core may be unrepresentative and inconsistent.

3.41 SAMPLING

The core will be recovered in a single piece enclosed in its rubber sleeve. It must then be cut with a saw into lengths convenient for boxing. Shorter lengths may also be cut and sealed for core analysis. Cut ends of the core are sealed with rubber caps and wrapped with fiberglass tape. An upward pointing arrow to show core orientation is marked on each piece of core using drillers chalk.

Unless permission is given to split the sleeve, geological evaluation must be restricted to small samples extracted from cut ends.

Rubber sleeved cores which, to some extent, maintain reservoir saturation (see Paragraph 3.45) and pressures better than conventional cores are sometimes frozen and sealed before shipment.

3.42   ORIENTED CORES

These are treated exactly like conventional cores except that extra structural orientation information is available. When handling or breaking the core, take care to avoid disfiguring or obscuring the scribe marks.

3.43   PRESSURED CORES

When pressured cores are taken, it is common for a pressure coring service crew to be present at the wellsite to handle core recovery and packing. The logging geologist will be required only as a geological observer. However, if no oil company geological representative is present at the wellsite, the logging geologist may also be required to monitor and report correct core processing procedures.

3.44   LOGGING

Logging is similar to conventional coring but special emphasis is placed upon establishing and monitoring background levels of mud salinity and nitrate ion concentration so that invasion may be estimated.

When the sensors are installed, monitoring of mud density, flowrate and standpipe pressure and reporting of variations is also of extra importance in controlling filtrate invasion of the core (see Paragraph 2.14).

3.45   RECOVERY

After the sealed lower barrel assembly is removed from the string, it is taken to a work area where a pressure guage is attached and the recovered pressure measured to test that a good pressure has been maintained.

A nonfreezing gel is then pumped into the assembly, closing a valve to seal the inner from the outer barrel and flush mud and water from the space between.

When flushing is complete, nitrogen pressure, equal to the measured recovery pressure, is reapplied and the whole assembly packed in a freezing box with dry ice. Complete freezing of the core takes a minimum of five hours. When frozen, the inner barrel is removed, as quickly as possible, to prevent thawing and returned to the dry ice freezing box for a further period of at least one hour. Required freezing periods vary, of course, with ambient temperature.

The inner core barrel is then cut into convenient shipping lengths, and small samples are taken from the cut ends for geological and hydrocarbon evaluation. Pressure-tight metal caps are clamped onto the cut ends and labeled with the normal well length and orientation information.

Throughout the whole operation, the barrel and core must be carefully watched for signs of thawing. If this occurs, contained gas will be free to bleed-off from the core, resulting in changes in saturations and pressure. At the first sign of thawing, the barrel must be returned to the freezing box.

The core is shipped to a laboratory for analysis in insulated boxes filled with dry ice. Pressured core analysis requires different equipment and procedures from conventional core analysis and is rarely, if ever, performed at the wellsite.

3.46   EJECTOR CORES

Cores from a diamond core ejector bit are flushed, of unknown orientation and poorly-known depth. They have no special value as cores and are logged and processed in the same manner as drilled cuttings.

3.47   SIDEWALL CST* CORES

3.48 GUN SET-UP

After running wireline logs, decisions must be made regarding the sidewall coring program. These include

- Formations to be shot
- Number of shots
- Type of bullet and charge
- Type of wire fastener

These should be made by the wellsite geologist and/or the logging geologist, in conjunction with the wireline logging engineer.

The logging instructions for the well will provide guidelines for selection of shotpoints, but considerations of hole size and irregularity, formation hardness and misfire record must also be made.

3.49 Hole Size

The wire fasteners used to retrieve the core from the borehole wall are available in a variety of lengths. Since the gun is run centered in the hole, large diameter holes will require longer fasteners (see Figure 3-18) in order to reach and penetrate the borehole wall. However, if long fasteners are used in a small diameter hole, the bullet will penetrate too deeply and become trapped. Attempting to recover the bullet will result in the fasteners breaking and the bullet being lost.

Careful scrutiny of the caliper log for zones to be shot should be made in order to decide fastener length. Some zones may be so washed-out that no attempt to core should be made.

---

* CST = Chronological Sample Taker.

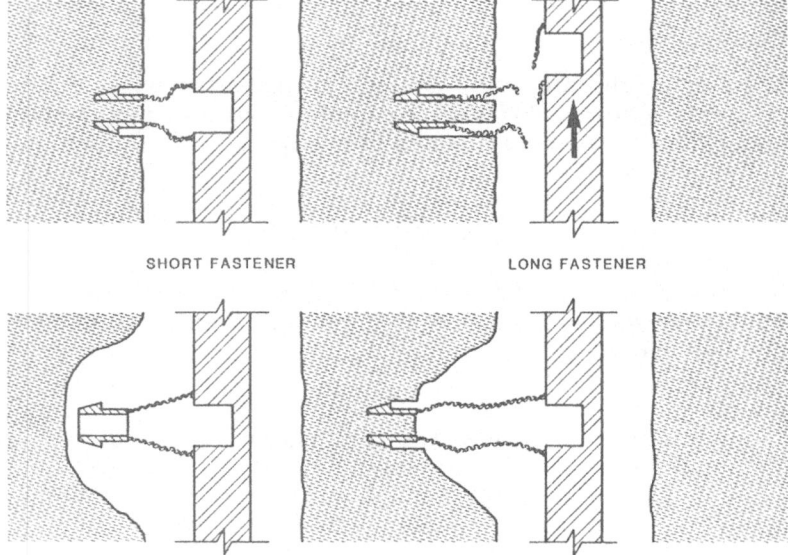

Figure 3-18. Selection of Correct Bullet Fasteners
for In-Gauge Hole and Overgauge Hole

3.50 Formation Hardness

Where formations are hard, it may be necessary to use specially hardened bullets or larger explosive charges to ensure sufficient penetration to obtain a good core. Softer formation bullets may fail to penetrate or be broken in the impact, causing the core to be lost.

On the other hand, hard formation bullets will penetrate too deeply into soft formations, become trapped and lost.

3.51 Misfires and Lost Bullets

Certain bullets may misfire or fail to fire. Others may fire, but due to the above considerations, may be lost or broken. The previous success rate of sidewall coring in the area and of the wireline service company must be considered when planning a sidewall coring program. The number of runs and shots per run should take into account the possible number of second shots required.

3.52 EXTRACTION

On retrieval of the gun from the hole, the sidewall gun is taken from the rig floor to the wireline unit work area.

FIRST, MISFIRED BULLETS AND CHARGES ARE REMOVED FROM THE GUN. NO CORE RECOVERY SHOULD BE ATTEMPTED UNTIL THIS IS DONE!

Removal of sidewall cores from the gun is normally performed by the wireline logging crew, but on extensive sidewall coring operations the logging geologist may be requested to assist or supervise them.

The fasteners are cut free of the gun with wire cutters and then unscrewed from the bullet. The base of the bullet is then removed and an extractor is used to push the core out of the bullet into a glass core bottle (see Figure 3-19). The bottle normally has a screw cap and a frosted label on which well and depth information can be written.

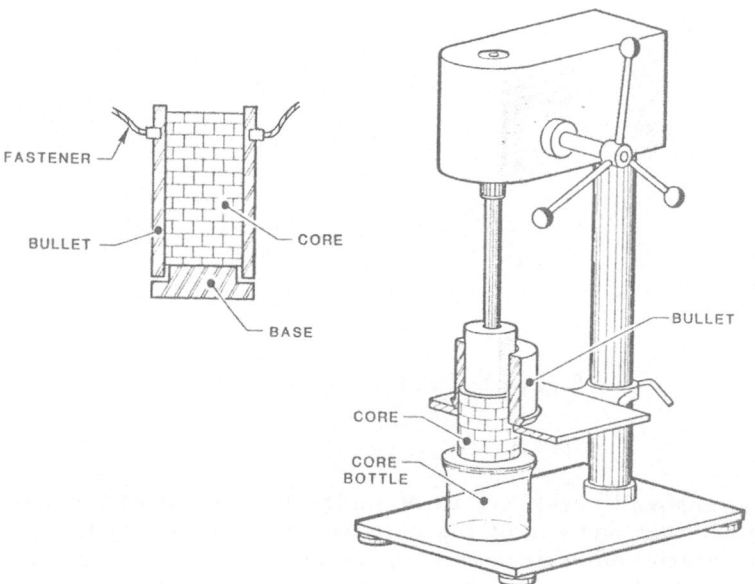

Figure 3-19. Removal of Sidewall Cores

Throughout this operation, great care is required to keep bullets, cores and bottles in the correct order!

3.53 CORE EVALUATION

3.54 Recovery

Estimating sidewall core recovery is important for financial reasons. Sidewall coring runs are charged for on the basis of the number of cores successfully shot. Prices are of the order of $60 to $80 per core.

The common criterion of successful recovery is a half inch. If less than a half-inch of core is obtained, not including filter cake, the core is not charged for. However, if this is the case, the wireline logging engineer will take the core and destroy it. If a further run with spare shots is not planned to allow the zone to be reshot, the geologist may decide that partial data is better than no data. In this case, a compromise is commonly negotiated with the logging engineer.

3.55 Lithology

Sidewall cores should be described carefully and a Core Report prepared for attachment to the bottom of the Formation Evaluation Log (see Figure 3-20). The report should include recovery and a tabulation of misfires and lost or broken bullets.

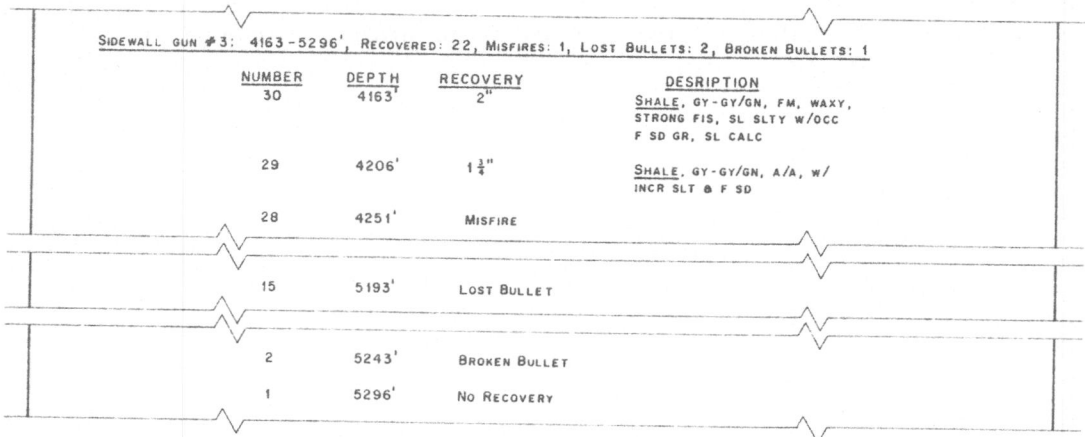

Figure 3-20. Sidewall Core Report

When examining sidewall cores, try to minimize the extent of breakage involved. Conversely, remember that one end of the core will be obscured by filter cake and the sides will consist of material pulverized and compacted by the bullet impact. Some scrapping of the core surface will be necessary in order to expose relatively representative material. For methods of more complete sidewall core description see Paragraph 4-42.

3.56 Hydrocarbons

Sidewall cores contain material which has been, at most, two inches from the borehole for some considerable time. They will be almost certainly invaded and flushed. Nevertheless, all cores should be tested for natural fluorescence. If any is seen it should be noted on the Core Report and a small piece of core (NOT THE WHOLE CORE) tested for cut type and fluorescence and these results added to the report.

Gas analyses from sidewall cores may be similarly unrepresentative; however, if requested, they may be performed in the same manner as conventional cores (see Paragraph 3.30). The sample may be obtained by punching a small hole in the metal cap of the core bottle.

3.57 Core Analysis

Sidewall cores are of little value in core analysis. Long exposure to the borehole greatly modifies their saturations and if they contain swelling clays, porosity and permeability.

The impact of the bullet pulverizes and compacts the outer part of the core while shattering and disaggregating the center. Porosities determined from sidewall cores are always unreliable and may be as much as five percent higher than conventional core plugs (that is, showing fifteen percent porosity compared to ten percent -- an error of fifty percent). Measured permeabilities will have even larger errors. Remember that permeability increases exponentially with porosity.

Nevertheless, core analysis is sometimes performed on sidewall cores. The whole core is required for this and no special sampling or processing is required. (See Paragraph 4.43.)

3.58 Logging

After all core runs have been completed, recovered and the Core Report prepared, the core depths must be noted on the log. This is down with small black triangles, centered in the core column, with the point to the right at the appropriate depth (see Figure 3-21). Normally it is only necessary to show cores which were recovered. However, if the oil company wishes that all cores attempted be noted on the log, non-recovered bullets may be denoted by hollow triangles.

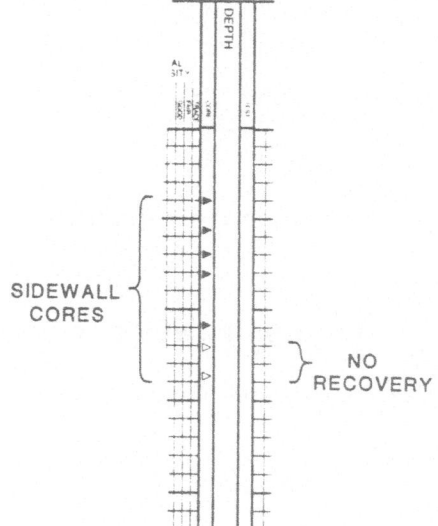

Figure 3-21. Sidewall Cores on the Formation Evaluation Log

3.59 SIDEWALL CORE SLICES

Core slices are almost always recovered broken into several pieces. The wireline logging crew will remove the pieces from the catcher and slide them into a cylindrical metal or plastic storage tube. If one tapered end of the core is contained in a piece large enough to remain vertical in the catcher, it may be possible to determine the orientation of the core, and this is marked on the tube with the core depth interval. The broken pieces may be reassembled and the core measured, described and reported like a conventional core (see Figure 3-12).

Core slices are noted on the log in the same way as conventional cores, but the lines marking the top and bottom of the cored interval slope toward each other within the core column (see Figure 3-22).

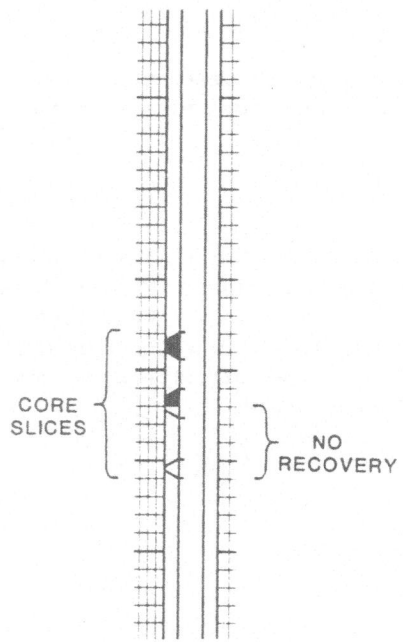

Figure 3-22. Core Slices on the Formation Evaluation Log

3.60   **ROTARY SIDEWALL CORES**

Rotary sidewall cores should be treated and logged in the same manner as other retrievable barrel cores.

4

CORE ANALYSIS METHODS

4.1   PRINCIPLES

4.2 POROSITY

The apparatus used to determine effective porosity is the Ruska Universal Porometer (Figure 4-1).

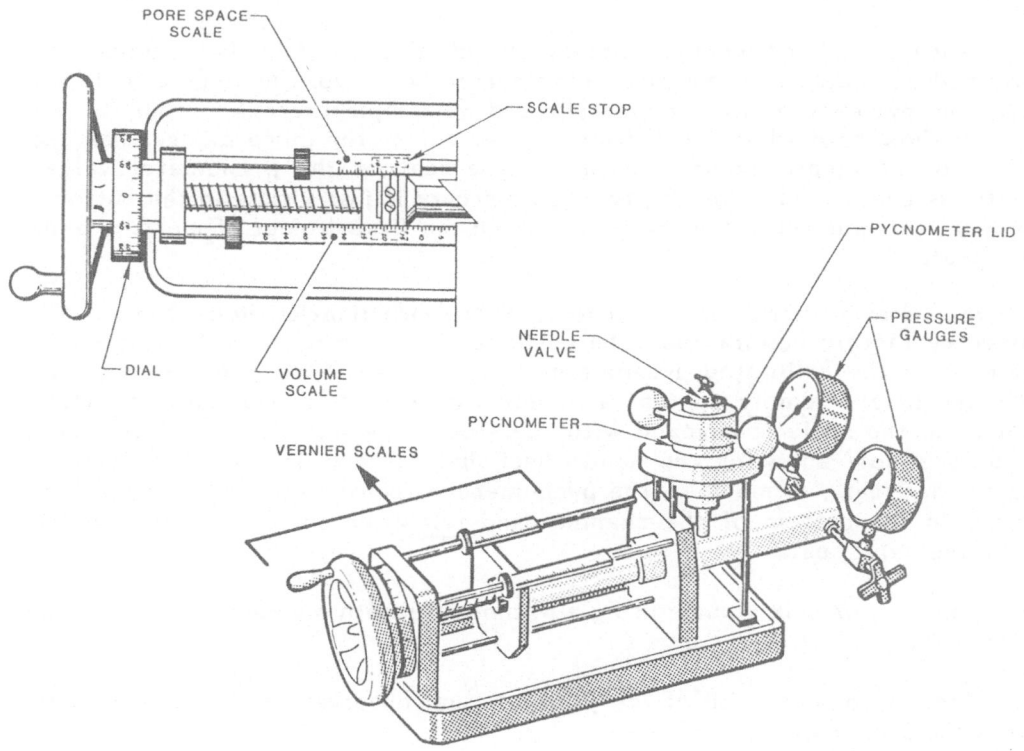

Figure 4-1. Ruska Porometer

The porometer consists of a 100 cc volumetric mercury pump to which a pycnometer is attached. The pump has a precision ground and honed, hard chromeplated cylinder, stainless steel plunger and an alloy steel measuring screw. The chamber of the stainless steel pycnometer has a volume of approximately 50 cc and admits cores up to 1-1/4 inches long and 1-1/2 inches in diameter. The porometer is furnished with one or two test quality pressure gauges, with ranges dependent on the method of operation.

The pycnometer lid has a rapid-acting breech-lock closure with an O-ring seal. A needle valve in the lid opens the chamber to the atmosphere. The movement of the pump metering plunger is indicated on two scales. The right- and left-hand scales provide, respectively, decreasing and increasing readings with the forward stroke of the plunger. Both scales are graduated to read the plunger displacement in cubic centimeters. The handwheel dial is graduated in 0.01 cc subdivisions and permits estimation of plunger displacement to 0.001 cc. The right-hand scale, which has decreasing graduations, is used for all volume measurements during determination of porosity, by the Kobe method. Additionally, the right-hand scale is used to provide bulk volume readings when the mercury injection method of porosity measurement is used. The mercury injection method is to be used only at the specific request of the client.

The porosity samples have to be approximately 10 cc in volume and preferably have a regular shape -- a right cylinder or cube is desirable. Irregular shapes with hollows should be avoided, as air will be trapped in them and give erroneous results.

The porosity samples should also be thoroughly solvent-extracted and dried.

The volume scale and handwheel dial provide direct readings of the bulk volume of a solid or porous body, in cubic centimeters. When correctly set up, the volume scale will read zero with the pycnometer full of mercury and no sample present -- that is, the scale reading will show zero when the forward movement of the pump plunger has displaced mercury into the empty pycnometer up to the seat of the pycnometer valve. The pycnometer is considered to be empty when mercury is just visible in the bottom of the open pycnometer, and full when the lid is on and the first droplet of mercury appears in the valve seat.

The void volume of a closed, empty pycnometer is approximatley 50 cc; the exact volume is obtained by factory calibration. The yoke stop is set to read the factory calibration. For example, if the calibrated pycnometer volume is 47.15 cc, the yoke stop is preset with the pycnometer empty so that a volume scale reading equivalent to 47.15 cc is obtained when the scale is engaged with the stop. The instrument is then readied for operation by simply adjusting the hand-wheel dial to read 15, using the dial numbers slanting to the right. Remember, the pycnometer is considered empty when mercury is just visible in the bottom of the chamber and full when the first droplet of mercury appears in the valve seat.

The determination of bulk volume using a pycnometer with a volume of 47.15 cc consists of the following:

1. With the pycnometer lid off to permit visual observation, mercury is withdrawn until the pycnometer is empty.

2. The volume scale is engaged with the yoke stop and the handwheel is set to read 15, slanted right to conform to the calibrated volume of the pycnometer chamber.

3. The sample is placed into the pycnometer and the lid is locked in position, leaving the valve open.

4. Mercury is pumped into the chamber until a bead of mercury appears in the pycnometer valve seat.

5. The volume of the sample is read directly on the volume scale with hand-wheel dial numbers slanting right. For example, if the volume scale reading is somewhat greater than 12 cc and the dial indicates 30-1/2, the bulk volume of the sample is exactly 12.305 cc. Record bulk volume measurement (V_b).

To determine the effective porosity of a core sample, the method makes use of Boyle's Law to determine the actual grain volume of the sample which is then calculated with the bulk volume to give the porosity. The theory of the method is as follows.

Assume a reference volume (V_R) of air enclosed in the pycnometer at atmospheric pressure (P_A). If the pressure in the pycnometer is then increased to a reference value P_R, then the air will be compressed to a new volume V_c. Then, according to Boyle's Law,

$$P_A V_R = P_R V_C \tag{4-1}$$

If the pressure is now brought back to P_A, the volume will return to V_R. If the core sample with a grain volume of V_g is then enclosed in the pycnomter along with air at atmospheric pressure, the total volume of air V_1 in the pycnometer will be given by

$$V_1 = V_R - V_g \tag{4-2}$$

When the system is then compressed to the reference pressure P_R, the final volume V_F will be the volume of the compressed air V_2 plus the grain volume V_g of the sample.

$$\therefore V_2 = V_F - V_g \tag{4-3}$$

By applying Boyle's Law (Equation 4-1)

$$P_A (V_R - V_g) = P_R (V_F - V_g) \tag{4-4}$$

or

$$P_A V_R - P_A V_g = P_R V_F - P_R V_g \tag{4-5}$$

Substituting Equation (4-1) in (4-5),

$$P_R V_C - P_A V_g = P_R V_F - P_R V_g \tag{4-6}$$

$$\therefore V_g = \frac{P_R}{P_R - P_A} (V_F - V_c) \tag{4-7}$$

Equation (4-7) is the basic equation of the Kobe method of porosity determination. Although it can be used in such a form for each sample, calculations would be tedious and errors could mount depending on the accuracy of gauge and barometric pressure measurements. A simpler graphical method is available and is used in wellsite core analysis.

To determine the total porosity, the bulk volume of the test sample is first found using the porometer. The sample is then crushed to grain size, and the grain volume of the sample is found using a glass volumeter. The volume of the sample should be approximately 9 to 10 cc and should be dried and extracted as for the previous method.

4.3   PERMEABILITY

The Ruska gas permeameter is the instrument used for measuring the permeability to air of the core sample (Figure 4-2). The instrument consists of a core holder of a modified Fancher type for drilled samples, a metal bushing adapter for shaped samples mounted in sealing wax, and three flowmeters with a selector valve covering different ranges. A pressure gauge and pressure regulator are used to regulate the inlet pressure from zero to one atmosphere. The core holder has a thermometer for measuring the temperature of the air as it passes through the core so that viscosity corrections can be applied.

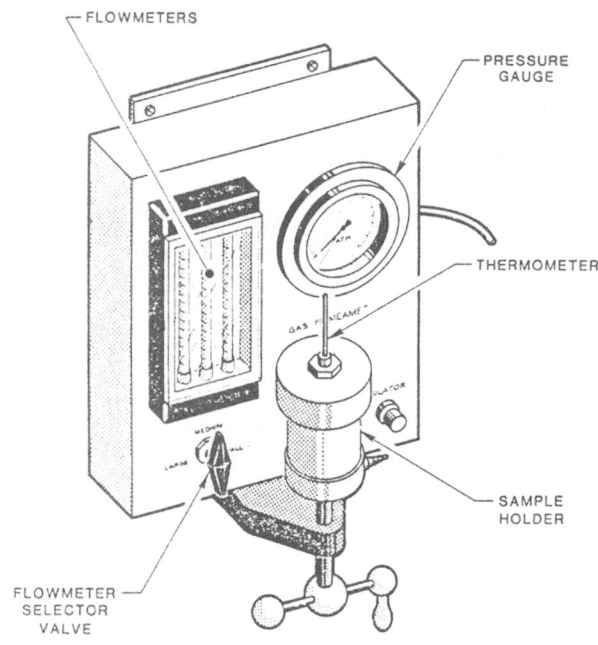

Figure 4-2. Ruska Permeameter

The core sample is sealed in the core holder in such a way that the air can pass only lengthwise through the sample (Figure 4-3). The desired air pressure is adjusted with the regulating valve and read on the pressure gauge. The flow through the sample is determined by the height of the center of the float in one of the flowmeters.

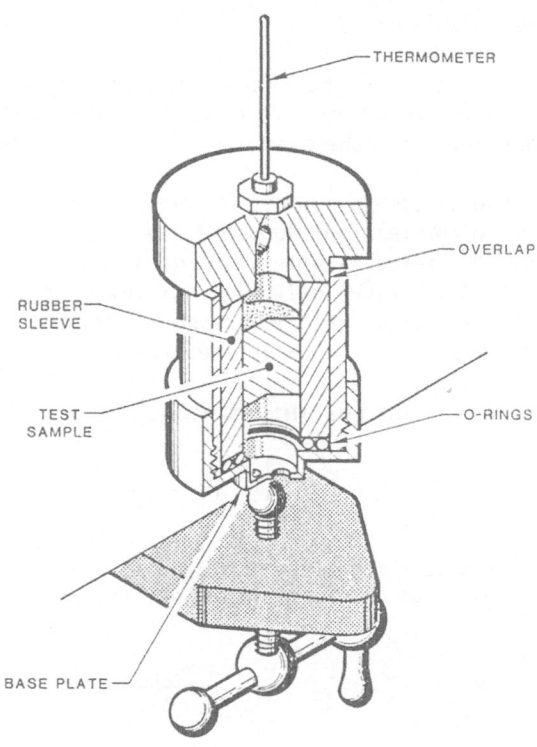

Figure 4-3. Permeameter Core Holder

Any compressed gas may be used to operate the Ruska permeameter. For convenience, air from the unit's regulated air supply is used for the permeability tests. The pressure of the compressed air should be between 20 and 40 psi, and the air should be clean and dry.

Core samples are prepared by cutting cylindrical sections from field cores with a chisel or knife, or with a hollow diamond core drill. It is advisable to standardize on one core size and to deviate from this standard only if special conditions make it necessary. A common size is 1/2 or 3/4 inch-diameter by 1-1/4 inch long. The orientation of the core sample with respect to the core should be noted. The permeability is standardly measured parallel to the bedding planes, and cylindrical plugs are drilled at right angles to the core axis. When cubic samples are prepared, the same sample may be used for determining permeability along three axes. Core sizes must be established accurately, since any inaccuracy in dimensions introduces errors.

Special core sizes may be necessary if the sample obtained from the field is too small or delicate for cutting to standard sizes, as will be the case with sidewall cores, or if the rate of flow through standard-size cores is not within the normal range of the instrument.

After the core section has been cut to size, all the contained water and oil must be extracted and the core thoroughly dried.

Delicate or fragile cores, which must be shaped by scraping and might be damaged or crushed when clamped in the core holder, are first imbedded with sealing wax into metal sleeves, which are then mounted into the holder.

When beginning the test, the large flowmeter is selected with a pressure differential of 0.25 atmospheres. If the flowmeter ball does not register between the 2.0 and 14.0 cm marks, then the medium flowmeter is selected and the pressure adjusted to read 0.5 atmospheres. Again, if the flowmeter ball does not register between 2.0 and 14.0 cm, then the small flowmeter is selected and the pressure adjusted to 1.0 atmospheres. The flowmeter is always selected before increasing the pressure.

The standard d'Arcy equation for permeability is used

$$K = \frac{Q \mu L}{A \Delta P} \qquad (1-9)$$

Since the downstream pressure is 1 atmosphere absolute, the pressure gradient is equal to the indicated gauge pressure. Thus the formula becomes:

Large flowmeter: $\qquad K = \dfrac{\overline{Q} \mu L}{A \; 0.25}$ or $\dfrac{4 \, \overline{Q} \mu L}{A}$ $\qquad\qquad$ (4-8)

Medium flowmeter: $\qquad K = \dfrac{\overline{Q} \mu L}{A \; 0.5}$ or $\dfrac{2 \, \overline{Q} \mu L}{A}$ $\qquad\qquad$ (4-9)

Small flowmeter: $\qquad K = \dfrac{\overline{Q} \mu L}{A}$ $\qquad\qquad$ (4-10)

where

$\quad$ K $\qquad$ = $\quad$ permeability, darcies

$\quad$ μ $\qquad$ = $\quad$ viscosity of the air, in cp (obtained from graph)

$\quad$ $\overline{Q}$ $\qquad$ = $\quad$ average rate of flow, in cc/sec (obtained from flowmeter reading and flowmeter calibration curve)

$\quad$ L $\qquad$ = $\quad$ length of sample, cm

$\quad$ A $\qquad$ = $\quad$ cross-sectional area of sample, cm2

$\quad$ ΔP $\qquad$ = $\quad$ pressure gradient, atm

The Ruska permeameter is calibrated at an upstream gauge pressure of

1.0 atmosphere for the small tube
0.5 atmosphere for the medium tube
0.25 atmosphere for the large tube

Sample Calculation

Observed: (1) Flowrator reading = 41.0 mm on the large tube

(2) Gauge reading = 0.25 atm

(3) Diameter of core = 1.9 cm (3/4")

(4) Length of core = 1.9 cm (3/4")

(5) Temperature = 20°C

From the curve of the large tube, 41.0 mm reading gives $\overline{Q}$ of 25.5 cm^3/sec. The viscosity of nitrogen at 20°C equals 0.0175 centipoise. The cross-sectional area (A) is calculated from the diameter of the cylindrical core and equals 2.83 cm^2.

Therefore,

$$K = \frac{4\,\overline{Q}\,\mu\,L}{A} = \frac{4 \times 0.0175 \times 25.5 \times 1.9}{2.83} \tag{4-8}$$

$$K = 1.202d \text{ or } 1,202 \text{ md}$$

4.4   SATURATION

Several methods are available for determining the fluid content of a core sample. The residual oil and water saturations are usually determined by one of two methods. The one most commonly employed by logging companies is that of retorting a sample in an electrically heated, water-cooled condenser-type still. It is the more rapid method for determining liquid content and is capable of producing reasonable, accurate results. A better but more difficult method is the extraction distillation method which is only very rarely used at the wellsite.

4.5   Retort Procedure

Figure 4-4 illustrates the basic components of a retort. The sample is contained in a stainless steel chamber, sometimes referred to as a test bomb, which has a tight-fitting screw top. The sample chamber sits inside the insulated electric heater, with its outlet stem making a vapor-tight seal with the top of the condensing tube. The condensing tube is fitted with an air fin-cooling section, and is further cooled by water circulating in a chamber around the tube. A calibrated glass receiving tube collects the liquids from the condensing tube.

98

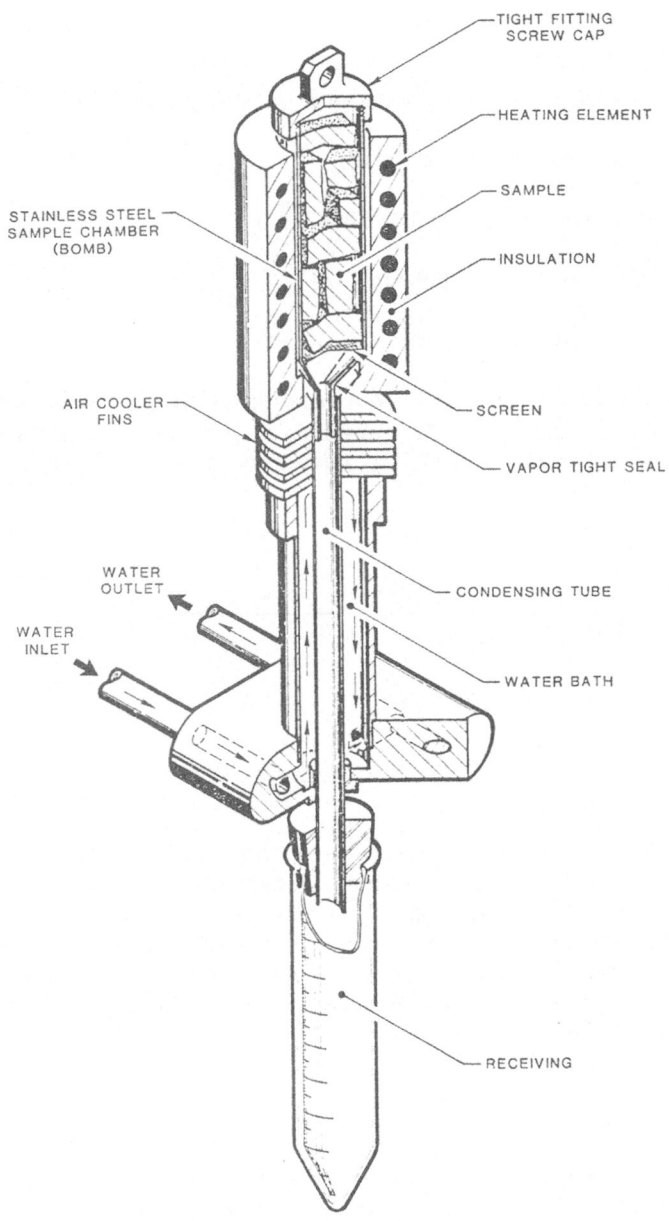

TIGHT FITTING
SCREW CAP

HEATING ELEMENT

SAMPLE

STAINLESS STEEL
SAMPLE CHAMBER
(BOMB)

INSULATION

SCREEN

VAPOR TIGHT SEAL

AIR COOLER
FINS

CONDENSING TUBE

WATER
OUTLET

WATER
INLET

WATER BATH

RECEIVING

Figure 4-4. Saturation Still Retort

A fragmented and accurately weighed core sample of about 100 grams is put into the sample chamber, and the top screwed on tightly. The sample chamber is then put into the retort which has been preheated to 1200°F, and the cooling water is turned on. Water will begin to collect within two to three minutes after the sample is placed into the heated still. A calibration graph is then plotted of retort water collected versus time in minutes (Figure 4-5). The resulting curve rises for the first 5 or 15 minutes while most of the free water is distilled. A plateau will eventually be reached (at 7-1/2 minutes in this example) which marks the end of the distillation of free water. Water collected after this time is the result of dehydration of the hydratable minerals. The water saturation reading is therefore taken at the "plateau value." The water is then turned off so that there is no flow through the water bath, and the bath drained. The sample is then left for a further 20 to 30 minutes to vaporize the heavier oil. After this time, the recovered oil volume is recorded. To ensure good separation of oil and water, the sample can be centrifuged or the collecting tube can be rubbed between the hands.

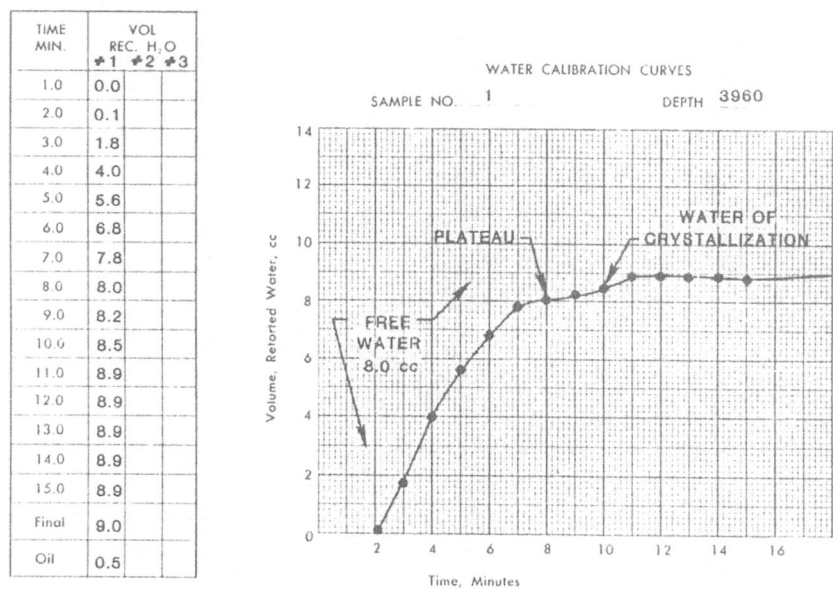

TIME MIN.	VOL REC. H$_2$O #1	#2	#3
1.0	0.0		
2.0	0.1		
3.0	1.8		
4.0	4.0		
5.0	5.6		
6.0	6.8		
7.0	7.8		
8.0	8.0		
9.0	8.2		
10.0	8.5		
11.0	8.9		
12.0	8.9		
13.0	8.9		
14.0	8.9		
15.0	8.9		
Final	9.0		
Oil	0.5		

Figure 4-5. Saturation Still Calibration Graph

A correction factor has to be applied for the volume of oil collected to compensate for vapor losses, coking, and cracking of the oil. This correction has been derived empirically from calibration tests made on each type of formation oil.

4.6   Extraction-Distillation Procedure

The other method, less commonly employed in the field, is that of extraction-distillation. It involves distilling the water from the core sample by refluxing it in a solvent which is immiscible with water but miscible with oil. The extraction of oil is completed by continued refluxing, after which the sample is dried. The oil volume is calculated from total weight loss minus the amount of water recovered. It is possibly a little more accurate than the retort method. However, it requires the use of delicate glassware, requires more time, and there is a greater possibility of operator error in making the measurements.

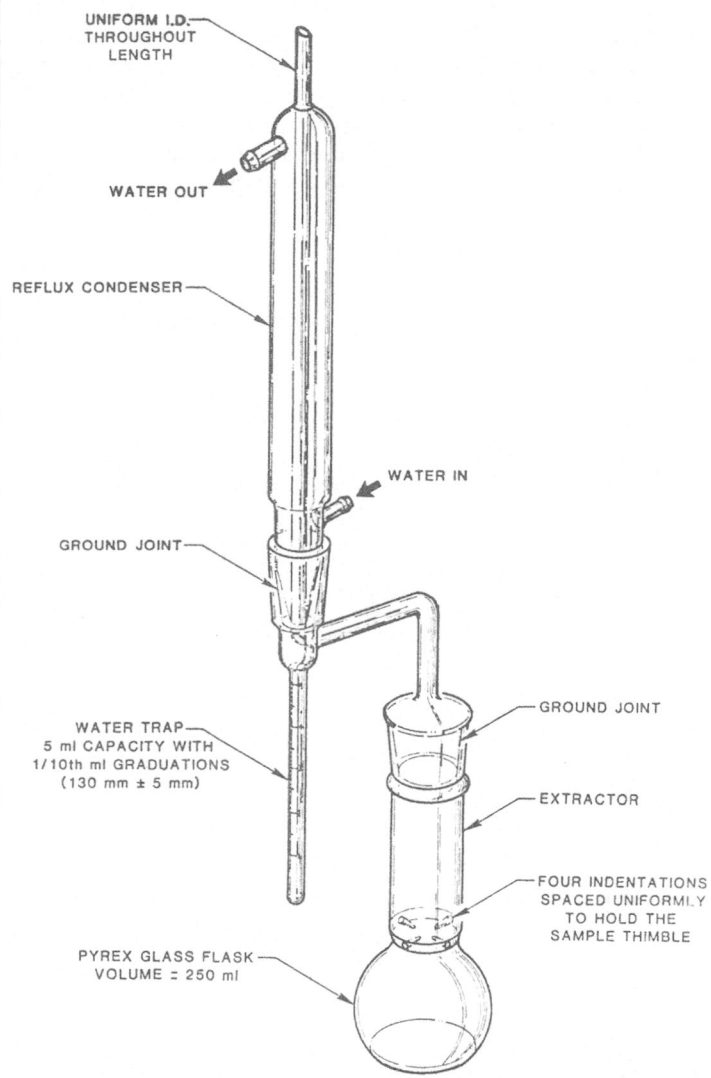

Figure 4-6. Extraction-Distillation Apparatus

Figure 4-6 illustrates the apparatus for the distillation extraction method and consists of a pyrex glass flask (or a metal still) heated by suitable means and provided with a reflux condenser which discharges into the graduated trap. The core samples are contained in round-bottomed aluminum thimbles.

4.7   SAMPLE PREPARATION

The wrapped and labeled samples are brought to the logging unit and laid out in order. They are then unwrapped, cleaned and core plugs obtained for core analysis. Samples should be unwrapped one at a time and carefully rewrapped afterwards. This will preserve them in the best possible condition in case resampling is necessary.

In some cases, it may be desirable to remove from a quarter to half an inch of the outside circumference of the core to eliminate some of the zone of drilling-fluid contamination. Always select the test samples for analysis from the interior of the core and a minimum of one-half inch from the outside circumference of the core. The outside one-half inch of the core will normally show the highest degree of alteration.

The porosity and permeability test samples must be selected immediately adjacent to the saturation sample, especially the porosity sample; the saturation results are directly related to and are dependent upon having the same porosity in the saturation sample as in the porosity sample.

4.8   SATURATION SAMPLE

The saturation sample requires a minimum of preparation. After selecting 100 to 120 grams of test sample from the centers of the cores, all that is necessary is to break the portion into 1/4- to 1/2-inch-size pieces just immediately prior to placing sample into the chamber for retorting. If the saturation tests are not run first, they must not be broken up into fragments but carefully wrapped whole in foil and stored in a refrigerator to prevent loss of fluids.

4.9   PERMEABILITY SAMPLE

The permeability plug may be cut with a diamond core drill if the core is firm to hard. This cutting method requires a fluid for cooling the core bit and removal of cuttings. The fluid must not cause any appreciable disintegration of the cementing material. For this reason, the air source at 20 psi is used in most cases to prevent overheating the core bit and sample. Only a small jet of air is required. Where the core is extremely hard, water may be used in place of air. Unconsolidated, friable samples may be shaped with a knife. The test plug must be extracted and dried prior to testing. If the test sample is fairly soft, whereby it could be compressed by the clamping action of the Fancher core holder, it must be mounted with sealing wax.

4.10   POROSITY SAMPLE

For hard to firm formations, the porosity test sample can be cut with the core drill in the same manner as the permeability sample or by breaking the core sample with a hammer into a single piece 9 to 10 cc in volume. For unconsolidated rock, the porosity sample is carefully shaped with a knife.

Care should be taken not to damage the clays when shaping the test sample, or a portion of the exterior pore openings may be sealed off. The test sample should not be too irregular or angular to prevent any trapping of air bubbles by the mercury when measuring bulk volume. After shaping, the porosity sample must be extracted and dried. Remember, the porosity sample must be selected from as near as possible to where the saturation sample was taken for pore volume comparison.

4.11 SAMPLE EXTRACTION AND DRYING

After cutting and sizing the porosity and permeability samples, all fluids must be removed before running tests. Oil and water may be removed either by extraction in the Soxhlet apparatus or in an evaporating dish with suitable solvent such as toluene or chlorothene. Upon completion of the extraction, the solvent is driven off by heating. Porosity and permeability samples may be extracted and dried together, but it is critical that such excessive heat is not used in drying that it will break down the hydratable clays, thus producing mineral alteration. Only the solvent and interstitial water must be removed in the drying process. The test samples should then be cooled and stored in a desiccator until ready for use. In the case of poorly consolidated samples, it is desirable to extract the fluids before final shaping.

WARNING

Be sure that the work area is
well ventilated, or do the extraction
in a well-ventilated safe area outside the unit.

4.12 Solvent Extraction

Extraction with an organic solvent is necessary to remove all hydrocarbons from the sample. It is recommended standard practice to extract all samples regardless of the presence or absence of visible shows. This provides for the removal of trace or residual hydrocarbons. It also gives consistency in results by consistent processing of all samples. When disposing used solvent, be sure to observe all wellsite safety rules and local environmental protection regulations. Guidance from the drilling supervisor or rig superintendent should be obtained when necessary.

In the case of low saturation with medium- to high-gravity oil, the samples may be extracted using the evaporating dish and solvent method. This is done by marking the samples for identification, placing them in an evaporating dish or pan, and heating the solvent to about 150°F. The container should have a tinfoil or regular cover. After heating for five to ten minutes, the container should be checked under the fluorescent light for color. The solvent should then be drained and a second application of solvent added to the samples and heated. This process should be repeated until all oil has been extracted from the samples and there is no fluorescence in the extraction solvent under the ultraviolet light.

If fluorescence examination indicates the sample is highly saturated or that the oil is medium to low gravity, the sample should be extracted by using the Soxhlet extraction apparatus and toluene (Figure 4-7). Fill the glass flask approximately two-thirds full of toluene, add a few 'antibump' beads (porous pot bead which prevents violent boiling and bumping), and turn the temperature-controlled hotplate on to boil the solution. The solvent fumes rise through tube A into the sample chamber and up into the condenser. When the solvent has condensed, it drips onto the sample and collects in the bottom of the sample chamber. When the level of the solvent in the sample chamber reaches the top of tube B, the sample chamber is drained of liquids by the siphon effect through tube B into the solvent flask underneath. The process then repeats itself. Fresh solvent is continuously cycled while the extracted hydrocarbons remain in the solvent flask. The sample is therefore repeatedly immersed in fresh hot solvent. The extraction process will continue as long as the hotplate is left on. The length of time that samples are left in the extractor will vary considerably. When the sample ceases to discolor, the solvent extraction is complete.

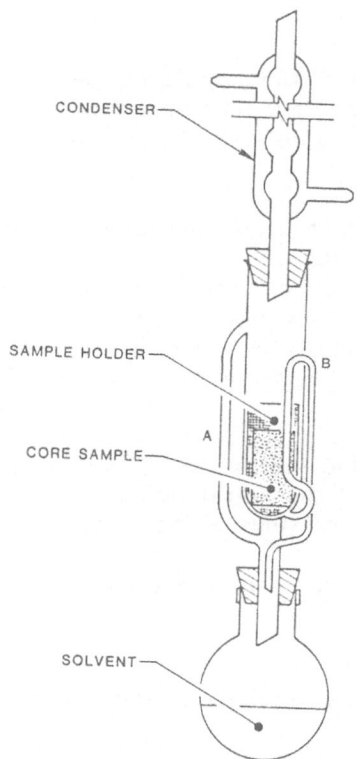

CONDENSER

SAMPLE HOLDER

CORE SAMPLE

SOLVENT

B

A

Figure 4-7. Solvent Extraction

4.13 Drying

Test samples should be dried in a drying oven with a vacuum fan. If this oven is not available, then the heat-lamp oven may be used. Tinfoil should be laid underneath the test samples. Drying samples in the oven, whether it has been extracted with solvent or not, should take at least twenty minutes for a friable sand and up to forty minutes for a tight, silty sand. The sample should be cooled for at least ten minutes in a drying oven with the heat element off. After cooling, the porosity sample should be weighed to the nearest 1/100 of a gram. This is the weight measurement for dry bulk density calculations and weight volume relationships on saturation pore volume. After being dried and cooled, the samples are ready for porosity and permeability tests.

4.14 SEALING THE PERMEABILITY SAMPLE IN WAX

Permeability samples need to be sealed in wax only if they are insufficiently consolidated to maintain shape when clamped in the permeameter sample holder. If this is the case, the following procedure is used.

Using the temperature-controlled hotplate, heat some sealing wax in a covered pot. Try to avoid boiling the water. Sprinkle fine sand on the counter and set a 38 mm metal sleeve on top of the thin layer of sand. Press down on the sleeve so that the level of sand inside the sleeve is about 1/4 inch.

Measure the length and diameter of the sample for permeability calculations and set the sample in the middle of the metal ring. The long axis of the sample should be perpendicular to the counter. Press the sample just slightly into the sand so the wax will not run over the end of the sample.

After the wax has been heated to the melting point, carefully pour the wax around the sample, filling the annulus between the sleeve and the sample only to the top of the sample. It is best to arrange to seal all the test samples which require wax treatment at one time, to save time in reheating and rearranging equipment. As soon as the wax hardens, the metal sleeve is ready to be placed in an R-48 rubber sleeve and inserted in the permeameter for analysis. Do not place it in the permeameter until the wax has completely hardened.

After analysis, the wax and sealed sample can be knocked out after slightly heating the metal sleeve. Clean the inside of the metal sleeve thoroughly, otherwise the wax on the next sample run will not stick to the sides of the sleeve.

The final preparation of test samples is explained under individual core analysis test procedures.

4.15 ANALYTICAL PROCEDURE

Core analysis requires that a number of tests be performed on several samples (see Figure 4-8). Quality and consistency of results is obtained by performing tests in a timely and efficient manner. The following procedures are designed to allow wellsite core analysis to be performed in the minimum of time with maximum efficiency and data quality. This is done by integrating the procedures of the three major tests, porosity, permeability and saturations, to minimize waiting time between stages.

4.16 PREPARATION

4.17 Long-Term

If core analysis is to be performed on the well, the core analysis set should be installed, tested and inventoried well in advance. Any malfunctions or deficiencies must be determined and corrected in time to allow core analysis to be performed correctly and reliably.

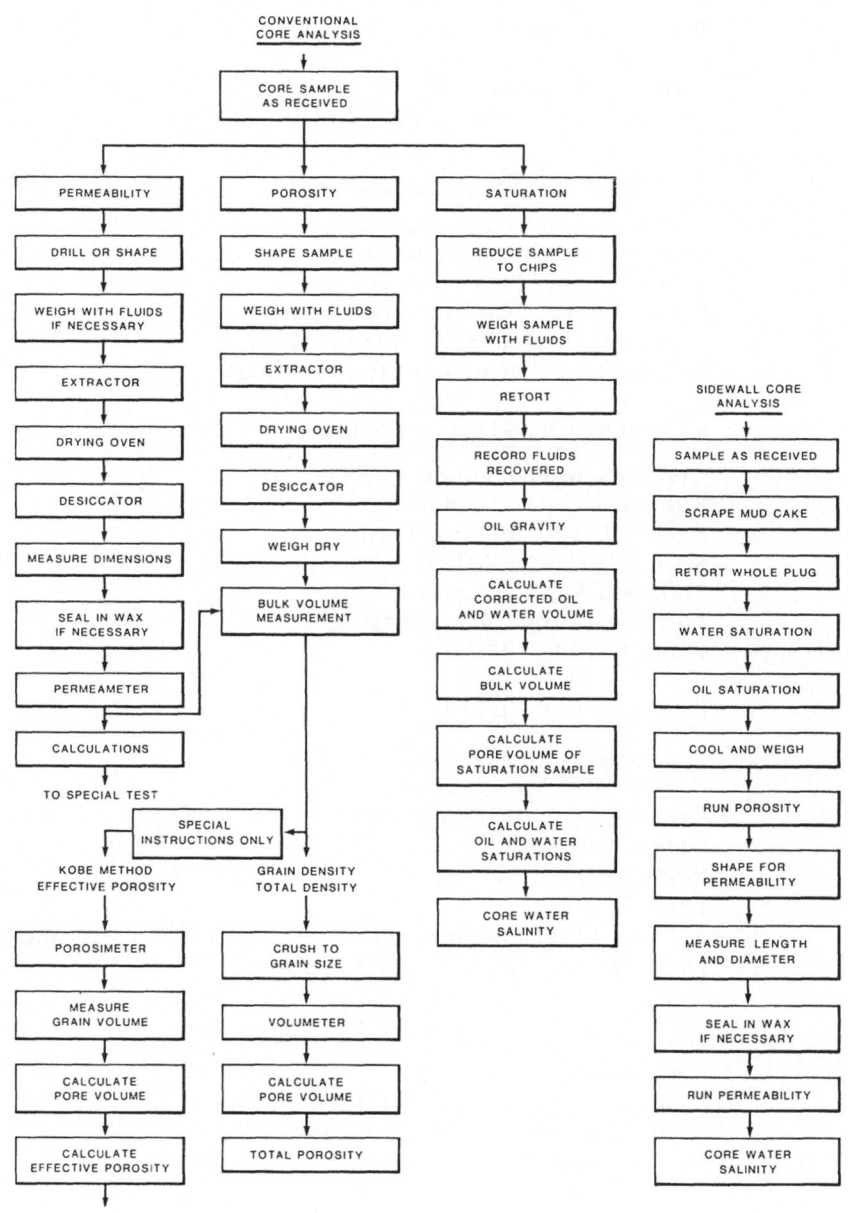

Figure 4-8. Core Analysis Procedures

The core analysis set should contain the following:

QUANTITY	ITEM
1	Ruska Permeameter with Wall Mounting Screws
1	Ruska Saturation Still Assembly (2 Stills) with Wall Mounting Screws
1	Ruska Porometer
1	Triple Beam Balance and Wooden Case

PERMEAMETER SUPPLIES

QUANTITY	ITEM
6	Stainless Steel Core Sleeves
2	Rubber Core Sleeves, R-20
1	Rubber Core Sleeves, R-48
1	Permeameter Thermometer, Spare
2	Core Holder Rubber O-ring, 2" O.D.
1	Set of Three Permeameter Test Plugs

SATURATION STILL SUPPLIES

QUANTITY	ITEM
6	Glass Receiving Tubes, Graduated
8	Rubber Stoppers, 2 hole
1	Retort Tube Cleaning Brush
12	Retort Gaskets, Asbestos
4	Stainless Steel Retorts
2	Electric Cords for Retorts
1	Tempil Stick, 1200°F

POROMETER SUPPLIES

QUANTITY	ITEM
10	Pycnometer Gaskets, Spare
1	Pycnometer Valve, Spare
1	Packing Booster, Spare
1	Spanner Wrench
1	Paint Brush, Small
1	Eye Dropper
1	Set of 2 Steel Porosimeter Calib. Plugs (5/8 x 1/2 and 5/8 x 1)
1	Jar Mercury, Triple Distilled

OTHER CORE ANALYSIS SUPPLIES

QUANTITY	ITEM
1	3/4" Diamond Core Bit
1	Hydrometer, 10° - 45° Gravity
1	Hydrometer, 45° - 90° Gravity
1	Vernier Caliper
1	Set of Tongs, 9"
1	Mortar and Pestle Set, 45L
2	Casserole Dishes w/Wooden Handle, 43-H
2	Boxes Dennison Sealing Wax, 1-lb Box (Not Paraffin Wax!)
1	Plastic Beaker, 100 ml
1	Plastic Beaker, 250 ml
1	Plastic Beaker, 600 ml

1	Plastic Funnel, 2"
1	Clamps, w/Vinylized Jaws
1	Dovetail Support Clamp
1	Screw Pinch Clamps
2	Pkg. Aluminum Foil, 12"
1	2 oz. bottle, Demulsifier Aerosol Solution in Plastic Bottle
1	Tygon Tube 3/8 Drain 48" long
1	Tygon Tube 3/8 Input with Fitting 36" long
1	Soxhlet Extraction Apparatus, Flask-250 ml Tube 40 ml Condenser 40 mm
1	Box Alconox Cleaner
1	Russel Volumeter, Glass
1	Hydrometer Cylinder, Glass
1	Ruska Porometer
1	Core Drill Unit and Press
1	Upright Rod for Hotplate
1	Gallon Toluene
1	Gallon Ethyl Alcohol
1	Metal Floor Clamp for Porometer
1	Permeameter Test Plug Calibration Chart
1	Ruska Porometer Manual
1	Ruska Permeameter Manual
1	Ruska Saturation Manual and Parts List
1	API Recommended Practice for Core Analysis Procedure
12	Core Analysis Supply Inventory List
2	Saturation Worksheets, Pad 50
2	Permeability Worksheets, Pad 50
2	Effective Porosity Worksheets, Pad 50
1	Total Porosity Worksheets, Pad 50
3	Core Analysis Report Form, Pad 50

4.18 Short-Term

All required equipment and supplies should be set up and ready before the core is re-trieved. Before starting core analysis work, be sure to dress adequately in coveralls and rubber-soled shoes, and remove watches, rings, etc., as mercury reacts with gold. Have all the necessary equipment and supplies at your disposal and ready for use. Have a set of Saturation, Effective Porosity, and Permeability Worksheets prepared and clipped together.

4.19 EQUIPMENT CALIBRATION

4.20 Saturation Stills

Plug in the stills in order for them to reach their operating temperature while you are preparing the other equipment and samples. The water should be off at this stage (Figure 4-9). The time required to reach operating temperature will vary with the rig electrical supply. Check the stills regularly, as overheating will result in melting of the polyflo tubing.

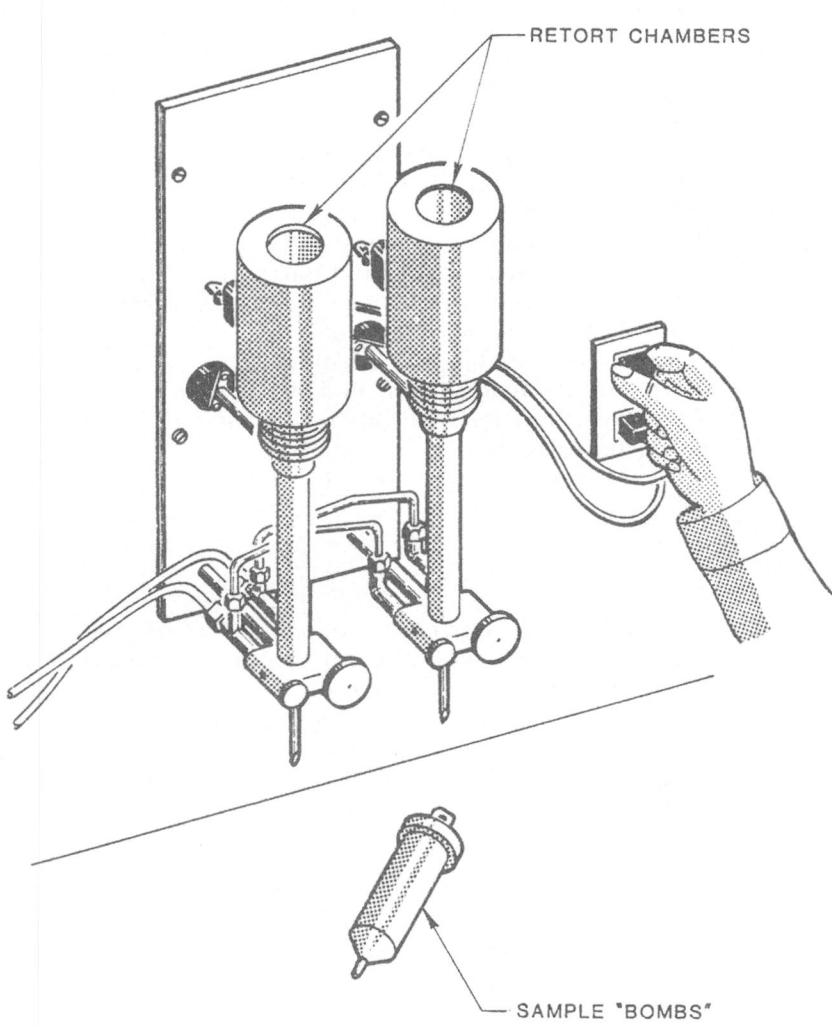

Figure 4-9. Plug in Stills

The collecting tubes should be clean, dry and have two drops of demulsifier in each. Put them in place, ensuring that the tubes are firmly on the rubber stoppers (Figure 4-10). Aside from preheating the stills to 1200°F before running a retort sample, no mechanical calibration of the stills is required.

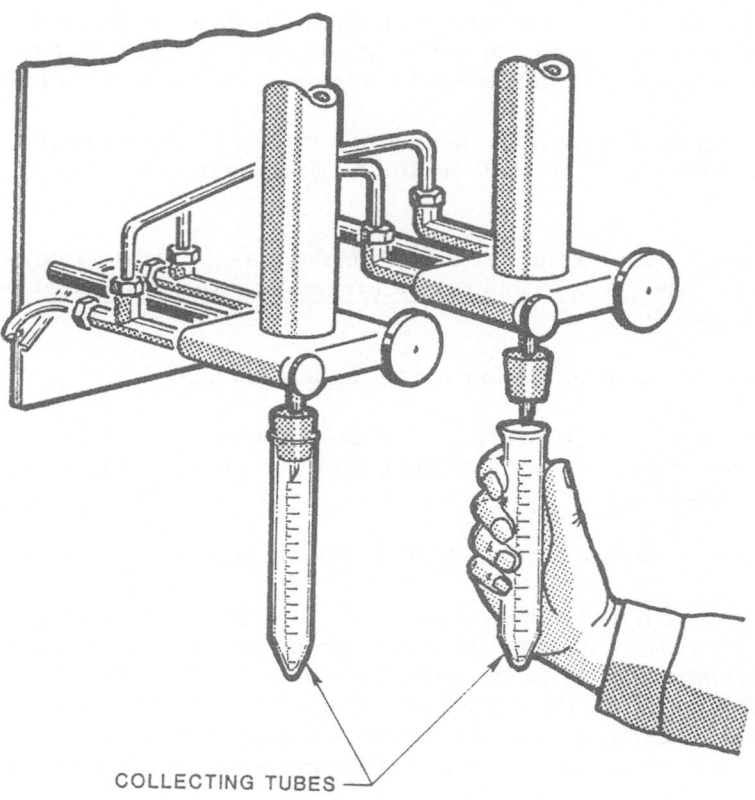

COLLECTING TUBES ─

Figure 4-10. Install Collecting Tubes

4.21 Porometer

Before measuring the porosity test samples, the porometer must be calibrated for proper operation. The procedure is as follows:

4.22 Mercury Level Check: A bead of mercury should appear at the bottom of the pycnometer chamber with the wheel crank back at the stop. Do not crank mercury back all the way into the pump, as this allows air into the system and upsets the calibration. Clean the chamber with a rag if there is any dirt in the chamber when empty. Add a few drops of mercury if necessary so that the bead appears when the crank is in stop position. Make sure that no air is trapped within the mercury volume.

4.23 Air Removal: At times, air may be drawn into the mercury pump. If so, the volume of air must be reduced to less than 0.5 cc at 750 psi for proper operation of the porometer. The actual factory correction factor is determined by Ruska. When all the air is removed from the pump the correction factor should theoretically be the same as the factory calibration. The procedure for removing the air is as follows:

a.   After charging the requisite amount of mercury, the pycnometer lid is locked in position, mercury is brought up to the pycnometer valve seat, and the valve is closed.

b.   Next, engage the pore space scale with the traveling yoke stop, and zero the hand-wheel dial. Always zero and take all readings with the hand-wheel hard over in the clockwise position.

c.   Make sure the valve to the low pressure gauge is closed.

d.   Increase the system pressure to 750 psi. The dial reading of the figures slanting left will be a direct measure of the correction factor. The correction factor should not exceed 0.50 cc.

e.   If the volume reading exceeds 0.50 cc, the air must be evacuated.

f.   The porometer is tilted backward until the face of the pressure gauge is horizontal. By alternately moving the plunger in and out of the cylinder, first pressure and then suction is applied to the gauge and system. Air trapped in the gauge, as well as in other parts of the system, is displaced by mercury and collects in a recess behind the packing gland. By bringing the porometer into its normal operating position, this air can be expelled into the pycnometer and subsequently released.

g.   The correction factor is then rechecked and the air evacuation process repeated until the volume of air is less than 0.50 cc. It may be necessary to repeat this procedure several times to reduce the volume of trapped air to an acceptable level.

4.24 Calibration: The method of calibration in the Ruska Porometer manual using simulated air volumes is not nearly as effective as the Exploration Logging system which uses steel plugs. The steel plug calibration system utilizes two different size stainless steel plugs (Figure 4-11). Bulk volume (V_b) and compression volume (V_f) are measured individually for the two plugs. A combined measurement is then made on the two plugs, giving three values for the grain volume calibration graph. The procedure is as follows:

a. Bulk volume measurements are taken on each of the two plugs and on the two plugs combined. Since the steel plugs have no porosity, their bulk volume is equal to their grain volume.

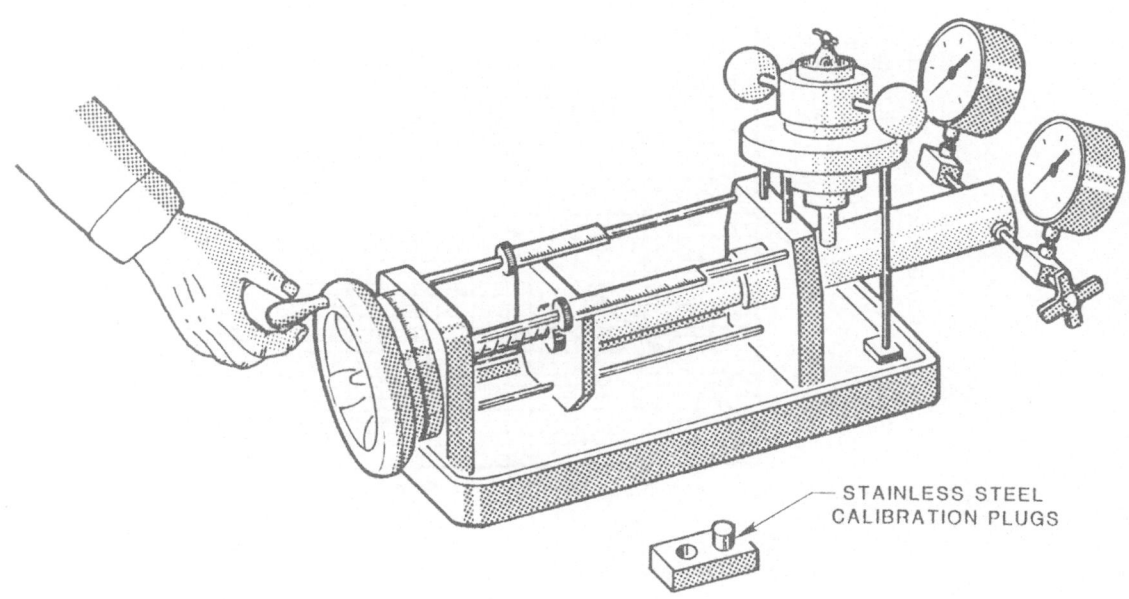

STAINLESS STEEL
CALIBRATION PLUGS

Figure 4-11. Calibrate Porometer with Calibration Plugs

b. Draw a graph on large-scale linear graph paper and plot bulk volume in cc (V_b) against the grain volume (V_g) for each of the three measurements. For accurate porosity measurements, the graph must be very accurately drafted using a sharp hard pencil (see Figure 4-12).

c. Obtain the exact volume of pycnometer chamber from the Ruska factory calibration sheet. Fill the pycnometer chamber with a volume of mercury whereby exactly 40.00 cc of reference air volume remains, and then place the first steel plug into the chamber.

d. Close the pycnometer valve and raise the system pressure to 30 psi reference pressure on the low pressure gauge. Record the final dial reading (V_f) on the calibration graph.

e.   Repeat the procedure for the second plug, and then for the two plugs combined. Plot the three V_f measurements on the left-edge scale of the calibration graph.

f.   The points on the calibration graph are the intersections of the grain volume and compression volume measurements. These results must give a straight line when connected or be remeasured until they do. It may also be necessary to prepare a new calibration graph after several hours due to changes in atmospheric pressure and temperature.

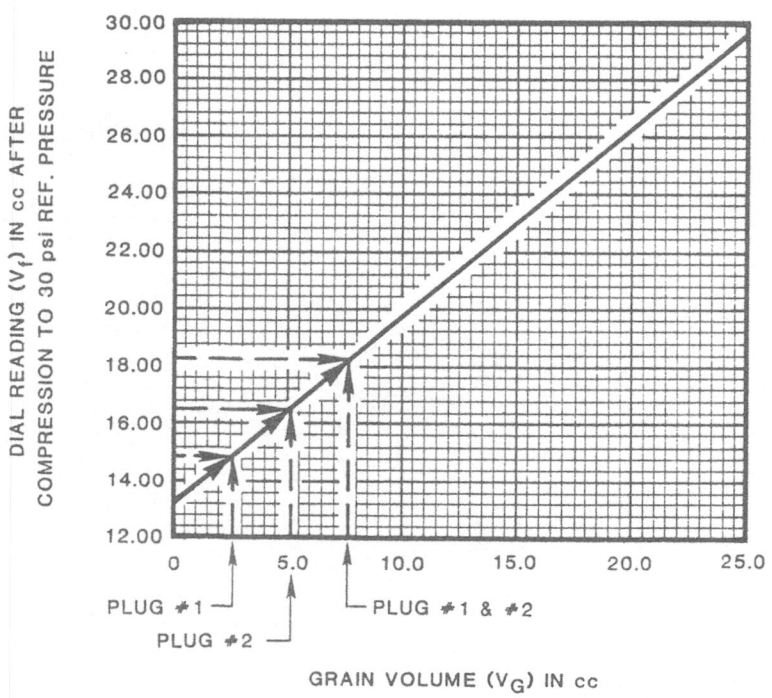

Figure 4-12.   Porometer Calibration Plot

4.25   Permeameter

Three test plugs are furnished with each permeameter. One plug is solid to check for leaks. The other two plugs have synthetic ruby orifices pressed in a brass holder. (A few test plugs are made of porous stainless steel. These are not quite as accurate as the plugs with fixed orifices.) The purpose of running test-plug analysis is to check the permeameter for leaks around the core test sample and for internal leaks or constrictions within the permeameter.

The most critical calibration test is the solid plug used to check for leaks. All three flowmeters must be checked for leaks using the solid plug for 1-atmosphere pressure. Minute leaks in the flowmeters and around the gasket seal can significantly affect permeameter measurements.

4.26 Solid Plug Calibration Test: (Figure 4-13)

a. Place the solid test plug in the R-20 rubber sleeve.

b. Make certain the rubber O-ring washer is in the bottom of the core holder. Place the R-20 sleeve and test plug into the core holder; tighten down the clamp.

c. Raise pressure to 1.0 atmosphere with the selector valve on the 'large' flow-meter. Check for any flow on the large flowmeter. Change to the medium and then to the small flowmeter, checking for any flow on each. There should be zero readings on all three flowmeters for a proper leak test; otherwise, the permeameter is not functional.

Figure 4-13. Permeameter Solid Plug Calibration

4.27 Flow Comparison Test: Flow tests are then made on the two orifice test plugs for comparison with the laboratory-measured values.

a. Follow Steps a and b from Paragraph 4.26 for sealing the test plugs in the core holder.

b. Measure the flow through the test plug in millimeters at 0.25, 0.50, and 1.0 atmosphere on the appropriate flowmeters.

c. Calculate the flow in cc/sec for each of the test plugs by using the Ruska outflow rate graph in the Ruska Permeameter manual. Figure 4-14 is an example.

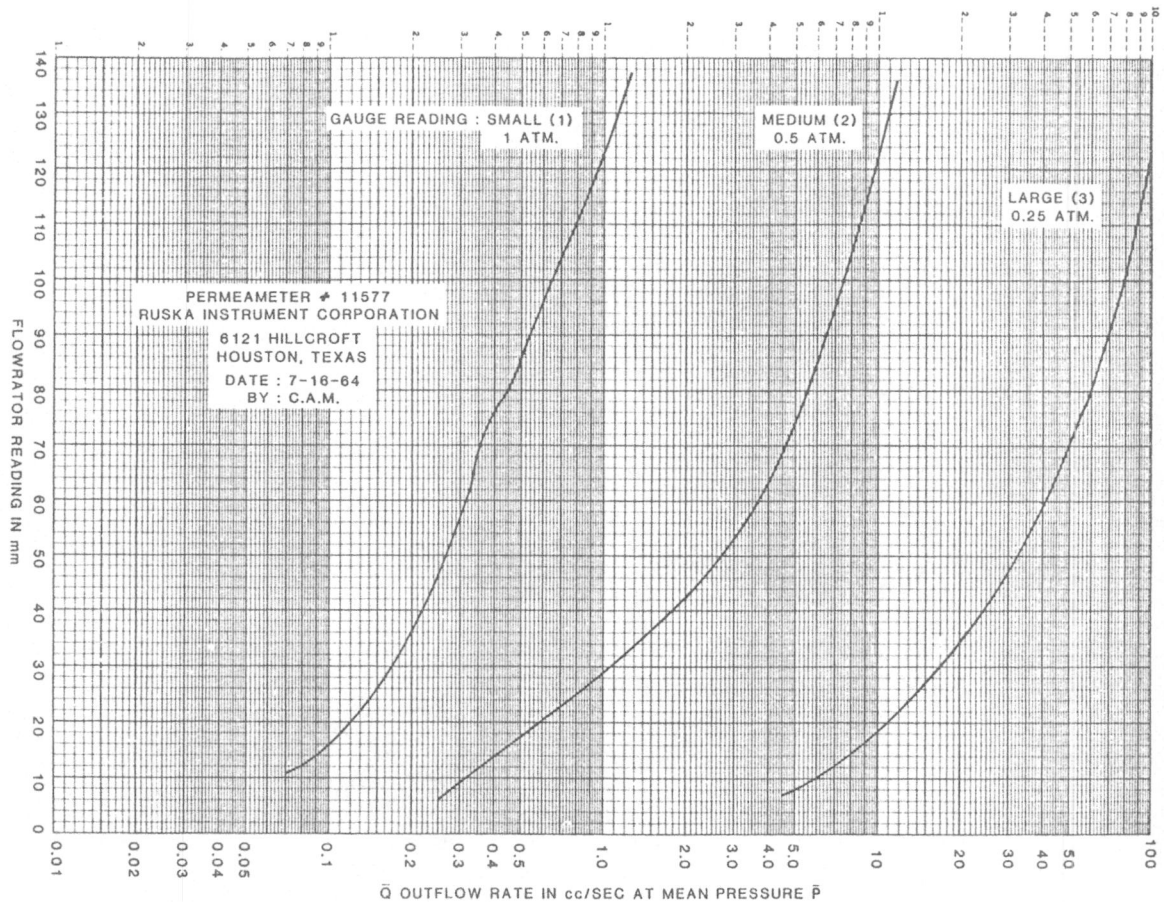

Figure 4-14. Permeameter Outflow Rate Graph (Example Only)

NOTE

Each permeameter has its own calibration graph which is specific to itself, and is identified by the serial number of the instrument.

d.   Determine the temperature-corrected viscosity of the air by using the "Gas Viscosity Graph" (Figure 4-15).

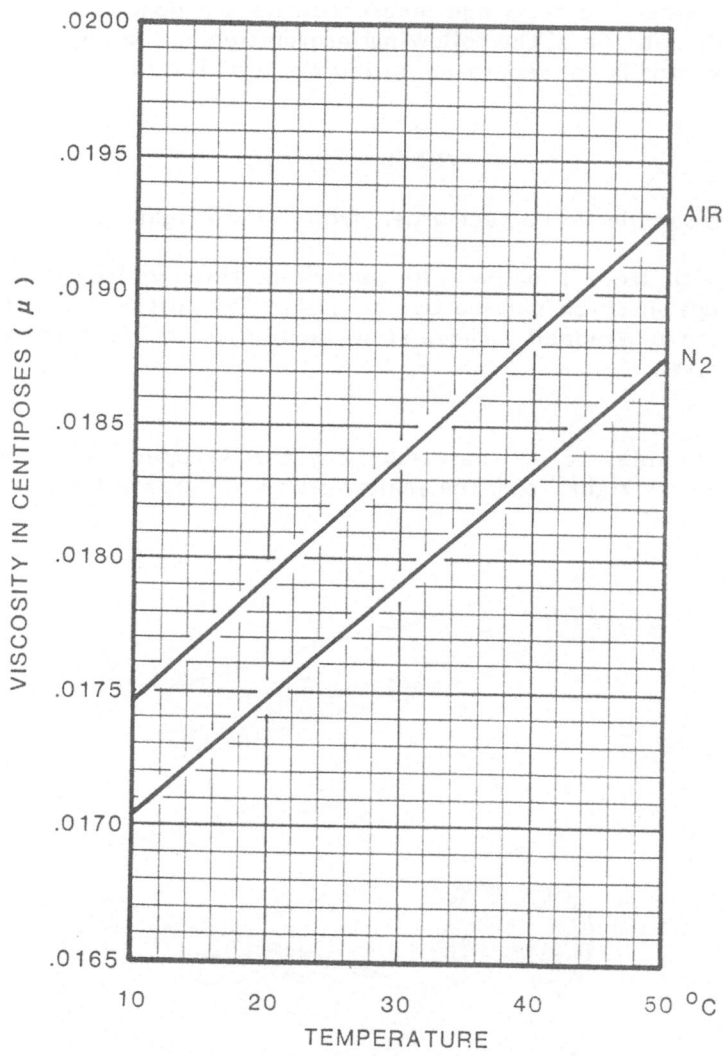

Figure 4-15. Gas Viscosity Graph

e.   Calculate: airflow (in cc), times corrected air viscosity (in cp), times flow pressure (in atm), and compare this figure with the calibration value. The calibrated value is furnished for all test plugs. Flowrates should check within 10 percent. This is approximately +/- 1.0 mm on the small flowmeter and 0.5 mm on the medium and large flowmeters.

4.28 CORE ANALYSIS TESTS

The procedures for testing saturation and permeability and for measuring porosity, are given in 63 steps. Since the tests and measurements are done 'simultaneously,' timing becomes very important. Thus, the following information is a listing of steps to follow to accomplish all three and is not separated into procedures for each test or measurement.

NOTE

Always keep samples in correct order, from left to right.

Beginning with Item 1 below and continuing through Item 8, the included steps must be carried out as quickly as possible once the core has been unwrapped. These steps pertain to the Porosity and Saturation tests.

1.   Line up the samples to be analyzed. Lay the wrapped core samples out on the bench from left to right, from the shallowest to the greatest depth (Figure 4-16).

Figure 4-16. Line Up Samples in Order, Left to Right

2. Mark wrapped samples from 1 onward (number 1 on the left). Note the numbers and depths of the samples on the Saturation worksheet (Figure 4-17).

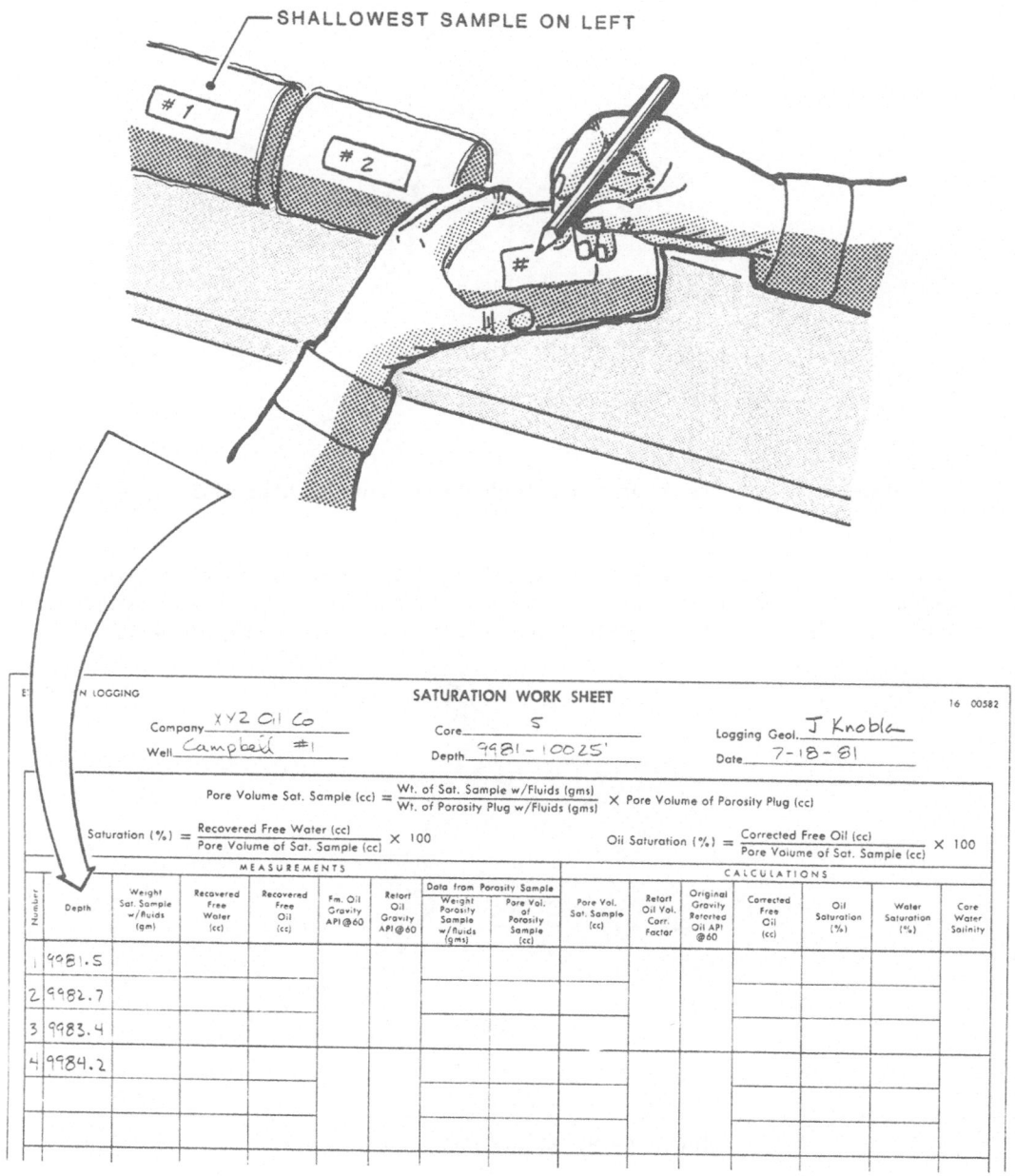

Figure 4-17. Number Samples and Record Depths

3. Unwrap sample number 1. Remove the core from its wrapping and break off a piece, full diameter and about 2 inches long, from the center of the core (Figure '-18). Avoid fractures, mud invasions and contaminant and try to get a representative sample for that depth -- taking several pieces if necessary. Rewrap the unused portions of the core and leave it in place.

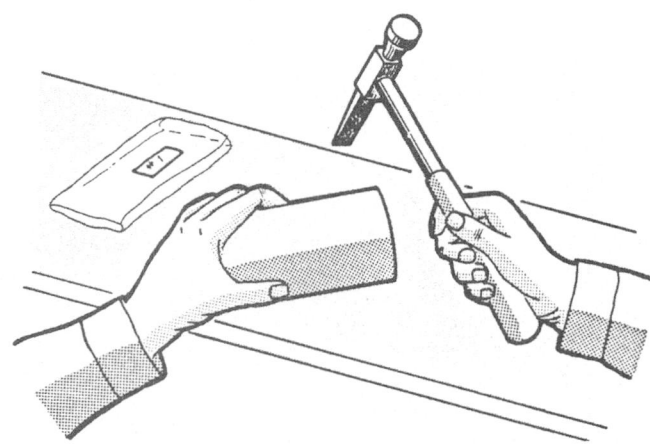

Figure 4-18. Break Out Two-Inch Piece from Center of Sample

4. Break off a suitable porosity sample, taking a piece from the center of the core of about 2 x 2 cm (20 g), avoiding rough edges or hollows as much as possible. Smooth out the sample to a cubic or cylindrical shape if necessary (Figure 4-19).

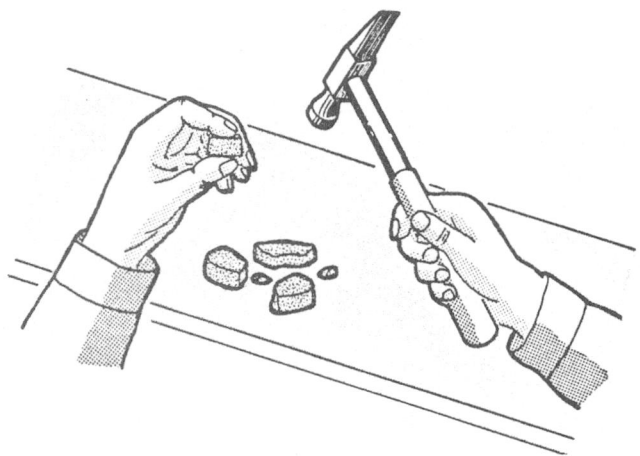

Figure 4-19. Shape Porosity Sample

For hard to firm formations, the porosity test sample can be cut with the core drill in the same manner as the permeability sample (see Item 13) or by breaking the core sample with a hammer to a single piece, 9 to 10 cc in volume. For unconsolidated rock, the porosity sample is carefully shaped with a knife.

Take care not to damage the clays when shaping the test sample, or a portion of the exterior pore openings may be sealed off. The test sample should not be too irregular or angular, to prevent any trapping of air bubbles by the mercury when measuring bulk volume.

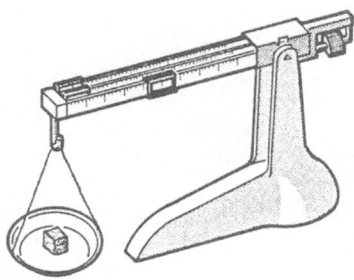

Figure 4-20. Weigh Porosity Sample

5. Carefully weigh this sample (Figure 4-20). Note down the sample number and <u>saturated</u> weight on the Effective Porosity worksheet. Mark the side of the sample with the appropriate number using a felt tip marker and place it in a marked sample tray or dish to be dealt with later (Figure 4-21).

6. Make brief Lithological and Show remarks on the worksheet.

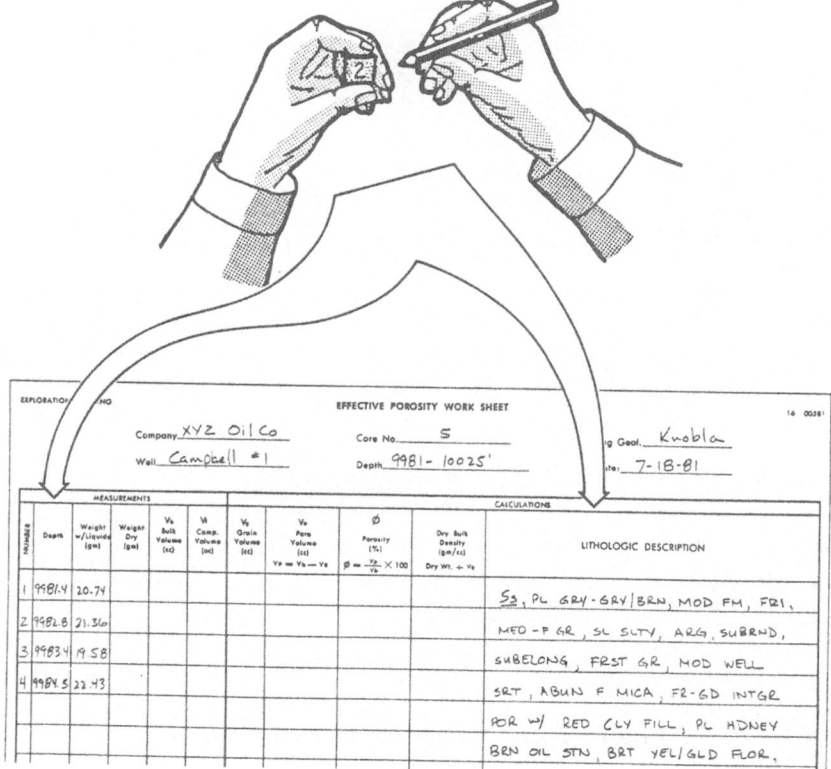

Figure 4-21. Record Depth, Weight and Lithology

7. From the center of the core, sufficient pieces of 1/4 to 1/2 inch in diameter should be broken off to make up to 80 to 100 grams (Figure 4-22). These should be weighed accurately to the nearest 0.01 gram, and <u>all</u> of the weighed sample must be loaded into the retort and must fit easily (Figure 4-23).

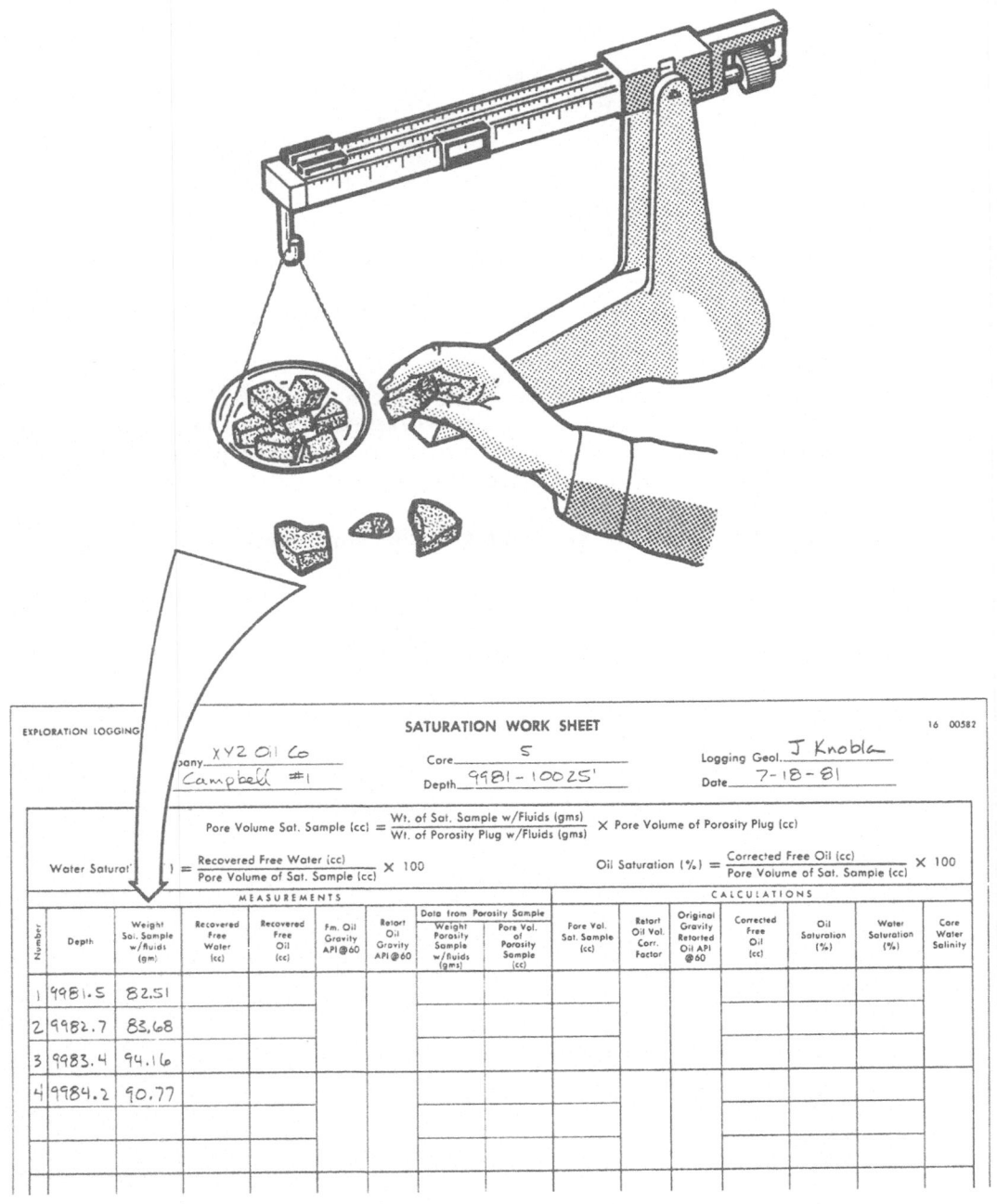

Figure 4-22. Weigh Saturation Sample

8. Choose a clean, dry, stainless-steel bomb (retort) and ensure that the asbestos gasket is in place in the lid. Screw the top on firmly and seal the outlet with masking tape. Tighten the top by hand with a bar or screwdriver through the lid. Mark the bomb with the appropriate number and lay it down in an appropriate place. Make sure the bombs cannot roll around.

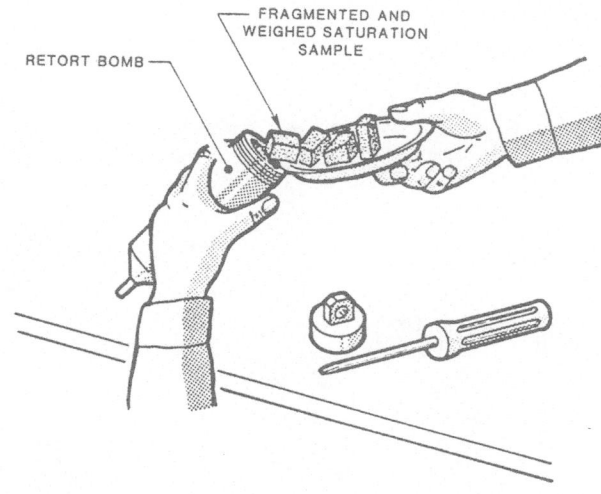

Figure 4-23. Load Retort

9. Wrap sample #1 and return it to its place. Clean up the work area, putting remaining core fragments into a labeled sample bag (Figure 4-24). Unwrap sample #2 and repeat the procedure as for sample #1, working rapidly with the unwrapped sample.

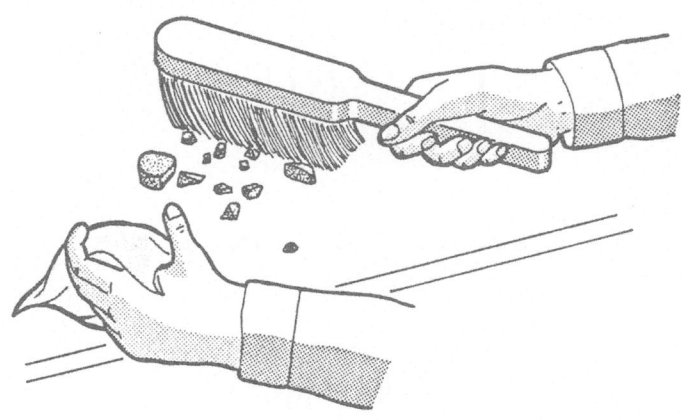

Figure 4-24. Clean Up Work Area

10. Load the retorts containing samples 1 and 2 into the stills, making sure that the masking tape has been removed from their ends. Make sure that sample #1 is on the left and that the collector tubes are in place and contain the demulsifier (Figure 4-25).

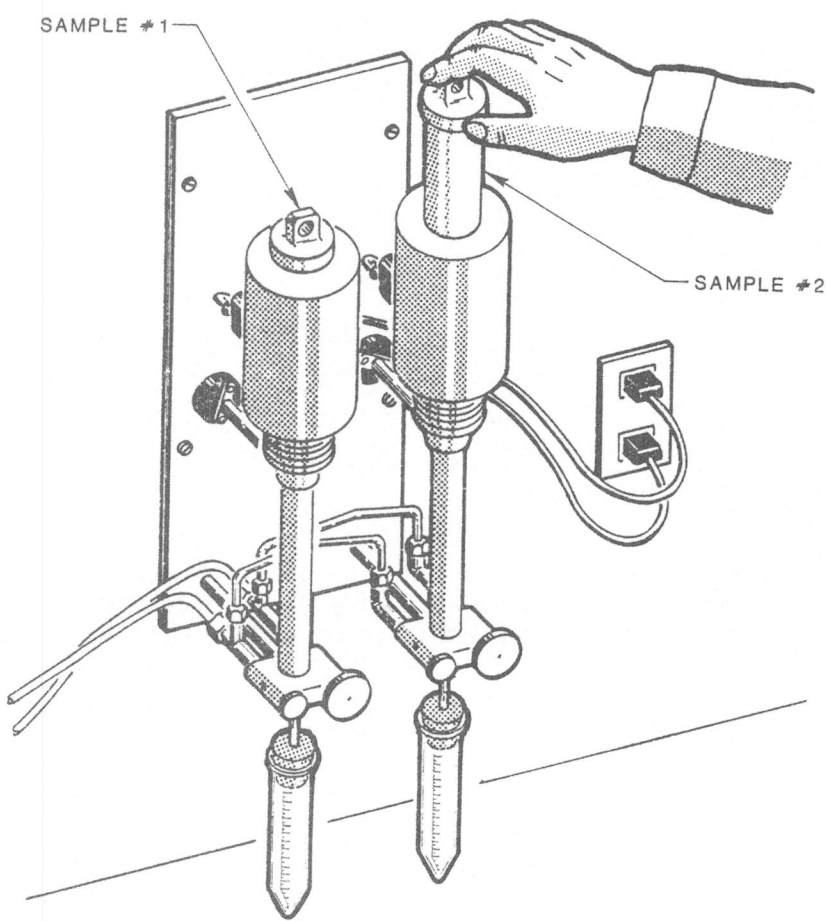

Figure 4-25. Load Retorts into Stills

WARNING

Be careful not to touch the
stills as they are extremely hot -- 1200°F!

11.   Turn on the water, pull out the left-hand knob on the stills (water inlet) and turn the right-hand knobs on the stills counterclockwise (full circulation), (Figure 4-26).

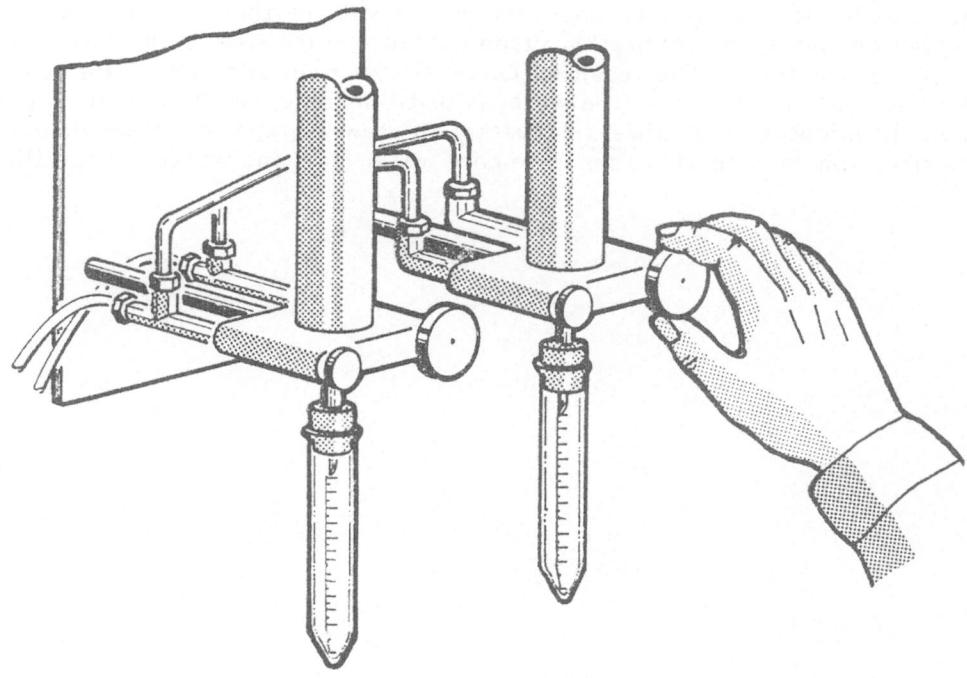

Figure 4-26. Turn Water On

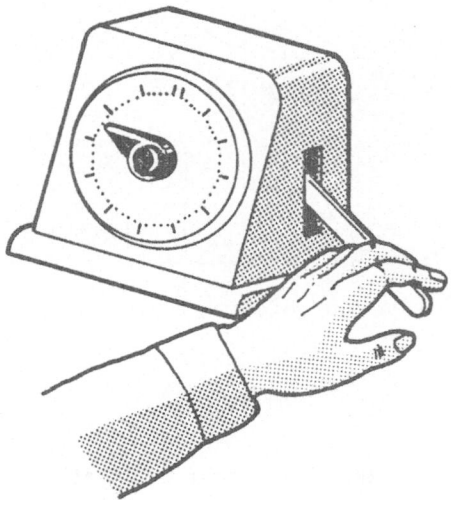

Figure 4-27. Set Timer

12. Set the timer for between 8 and 14 minutes, depending on initial verifications (usually for 12 minutes; Figure 4-27). Water will begin to collect within two or three minutes after the sample is placed into the heated still. Water volume readings should be recorded at one-minute intervals in the first instance and then plotted on the water calibration graph on the reverse side of the saturation worksheet (Figure 4-28). The resulting curve always rises sharply for the first 8 to 15 minutes while most of the free water is distilling, and usually reaches a plateau at about 10 minutes. It is <u>always</u> necessary to draw a graph of water-versus-time for the first sample at least, to separate core water from the water of crystallization.

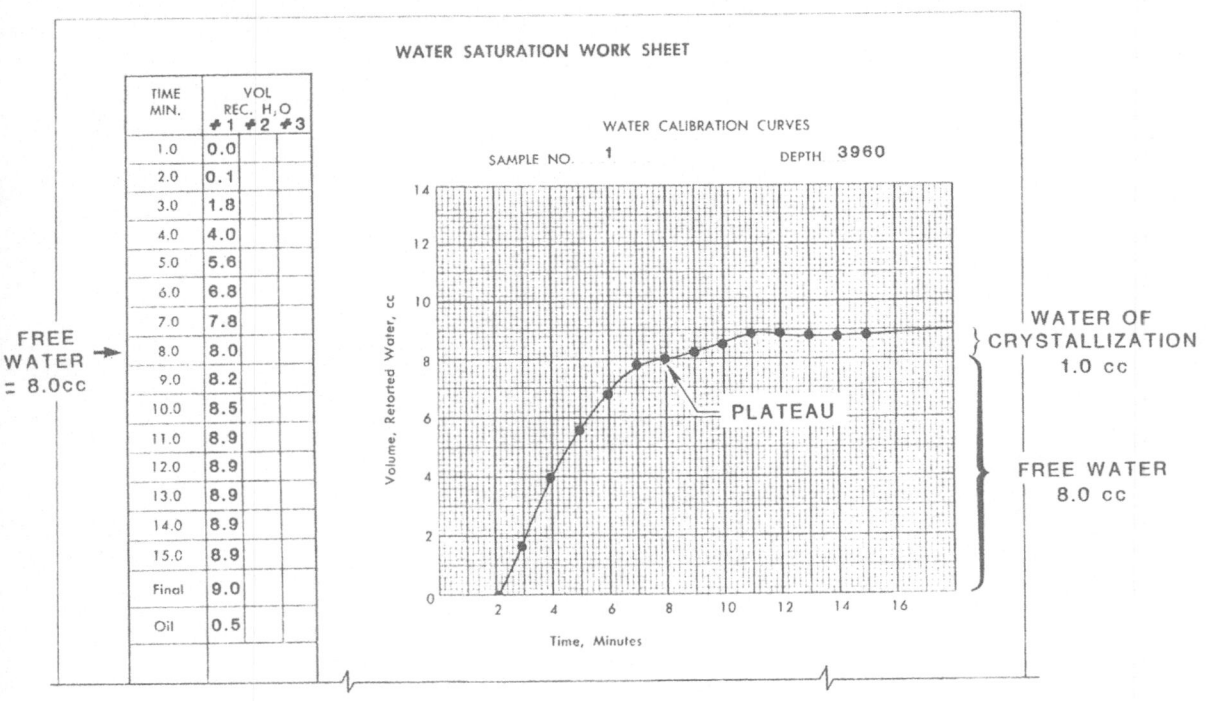

Figure 4-28. Plot Water Calibration Graph

After the first run the time to reach this plateau can be set on the timer, and it will not be necessary to record the water level every minute. A sheet of paper can be placed behind the tubes for easy reading (Figure 4-29).

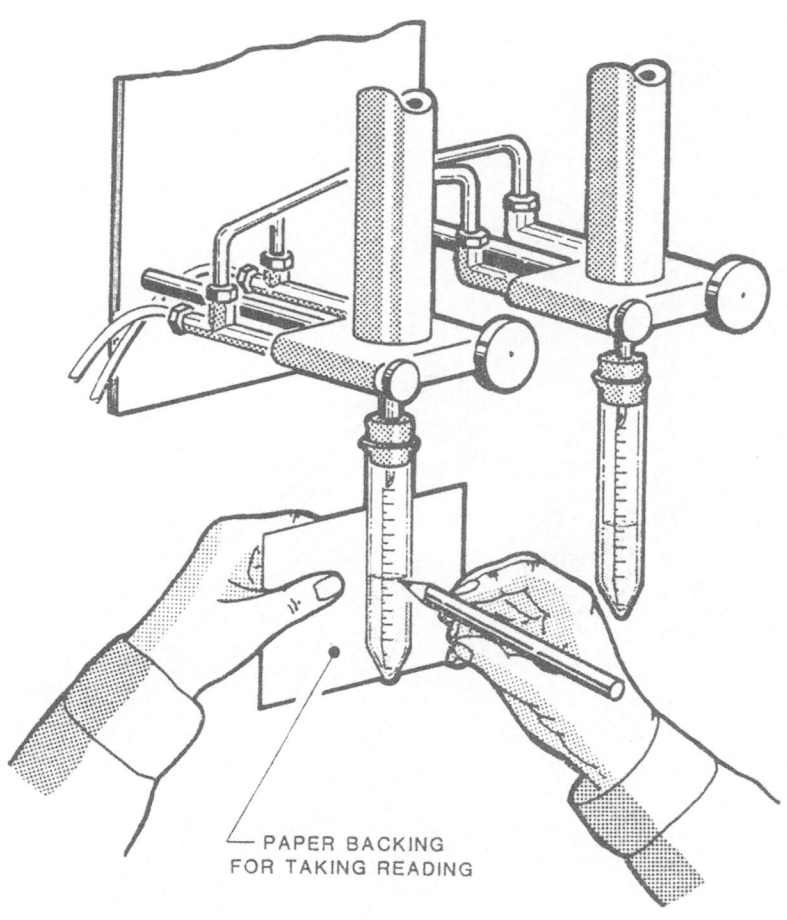

PAPER BACKING
FOR TAKING READING

Figure 4-29. Reading Collection Tubes

13. While waiting for the water readings from the retorts, there will be time to select and shape the permeability plugs from sample 1 and 2. The permeability plugs may be cut with a diamond core drill if the core sample is firm to hard. This cutting method requires a fluid for cooling the core bit and removing the cuttings. The fluid must not cause any appreciable disintegration of the cementing material. For this reason, the air source at 20 psi is used in most cases to prevent overheating of the core bit and sample. Only a small jet of air is required. Where the core is extremely hard, water may be used in place of air (Figure 4-30). Unconsolidated, friable samples may be shaped with a knife.

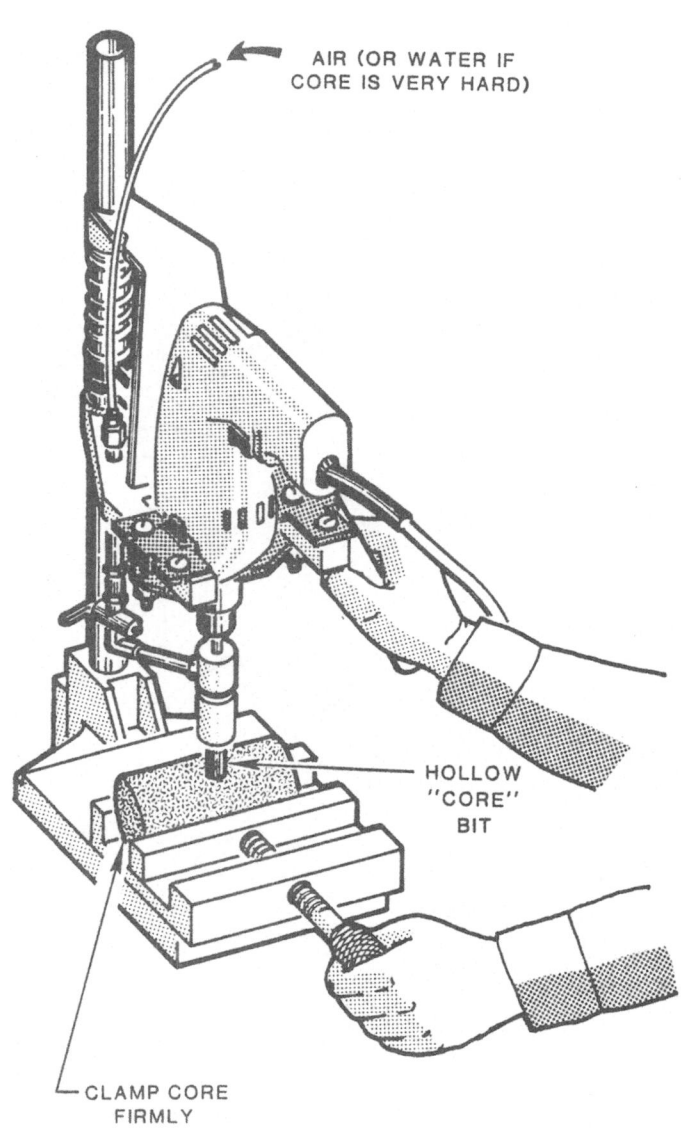

Figure 4-30. Cut Permeability Sample Plugs

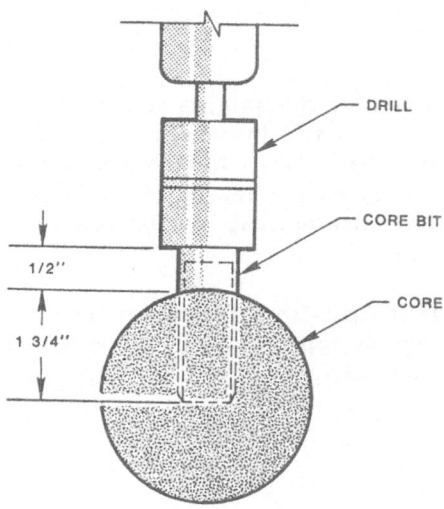

Figure 4-31. Cutting Plug

Consolidated Core Sample: Plugs for the permeability test should be drilled from the core sample to within about 1/2-inch from the upper end of the core bit (Figure 4-31). Upon retraction of the core bit, the plug is broken off using a small screwdriver. The permeability plug is always cut parallel to the bedding plane to determine horizontal permeability. If no bedding is apparent, the plug should be cut perpendicular to the wall of the core. A ring incision is then cut with a knife or hacksaw around the circumference of the upper end of the plug, leaving about 1-1/4 inches of plug length. With a pair of pliers, a knife, or a chisel placed against the incision, the end of the plug is broken off, leaving a fresh fracture at both ends of the plug (Figure 4-32).

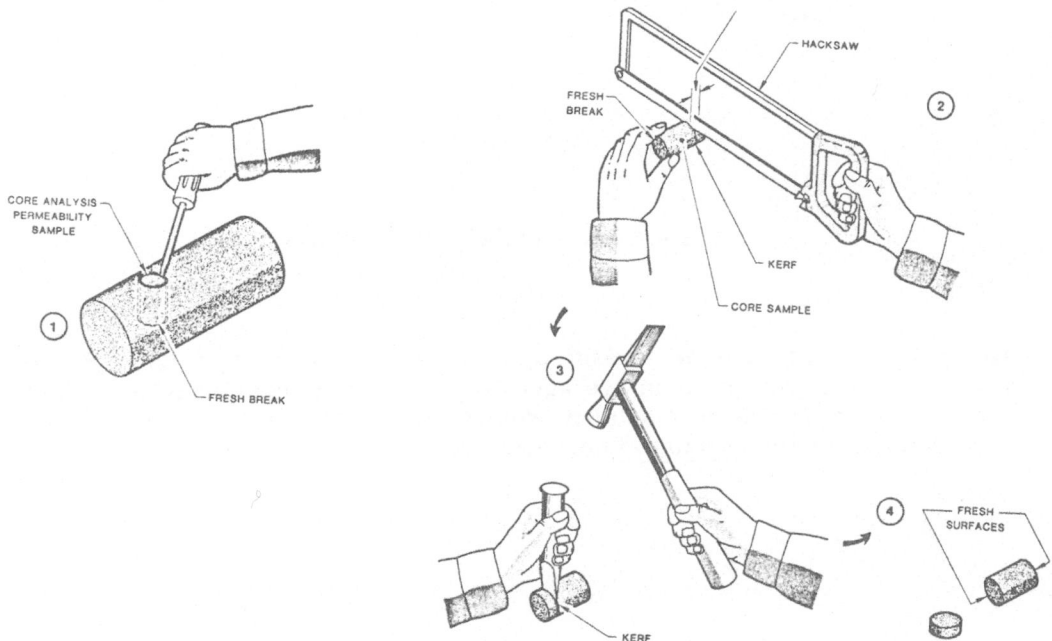

Figure 4-32. Preparing Sample Plug

● Unconsolidated Sample: If the sample is too friable to be cut with a core drill, it must be shaped with a knife into cylindrical or rectangular form. Again, the ends of the plug must have fresh breaks and not be abraded with the knife. If the test sample is fairly soft whereby it could be compressed by the clamping action of the Fancher core holder, it must later be mounted with sealing wax as described in Steps 49-51.

14. Number the side of the plugs with a felt tip marker, taking care not to mix up the samples, and place them in appropriate trays -- either with or separate from the porosity samples (Figure 4-33).

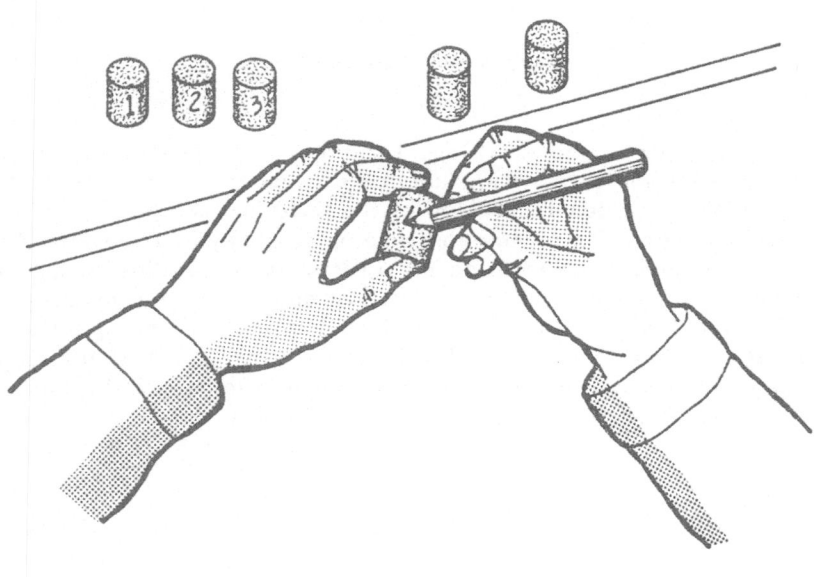

Figure 4-33. Number Sample Plugs

15. The plugs should then be measured, using Vernier calipers (Figure 4-34). Take several readings to obtain an average. Note down the mean length and mean cross-sectional area on the Permeability worksheet. Make <u>sure</u> that the plug is cut for measurement of the <u>horizontal</u> permeability.

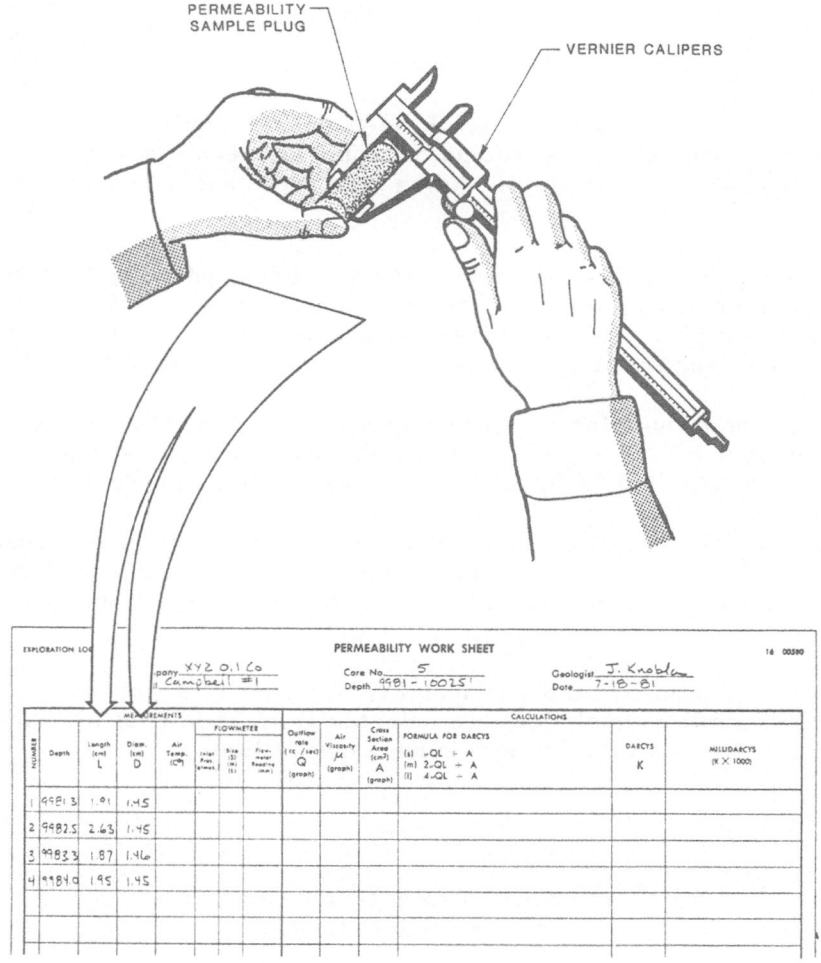

Figure 4-34. Measure Plugs and Record Length and Diameter

NOTE

For Saturation Test, retorts will probably
now be ready for water readings.

16. Note down the water readings on the Saturation worksheet. Take the plateau readings where the curve levels off (Figure 4-28).

17. Turn off the water by pushing in the left-hand knob on the bottom of the stills, turn the right-hand knob fully clockwise, and disconnect the water source. The stills should be full of water, but the flow should have stopped.

18. Set the timer for another 15 minutes, and, if present, turn on the extractor fan over the workbench.

19. Prepare clean retort bombs and collecting tubes, and clean the balance.

20. Unwrap samples 3 and 4 and repeat the foregoing, ensuring that all samples are clearly marked and ordered. Break-out and prepare as many samples as you have spare retort bombs. Make sure that the bombs cannot roll around, that they are in correct order, and that the orifices are taped over until they are to be used.

21. When the timer rings for the second time, drain the water off by pulling out the left-hand knob on the bottom of the stills and opening the right-hand knob fully counterclockwise. This will speed-up drainage of oil from the drain tubes.

22. Set the time for the final oil readings -- another 15 minutes. Heat the still to 1200°F to drive off the remaining heavy oil in the sample and condenser drain tube. A piece of aluminium foil can be placed on top of the still for additional heat.

After shaping, the porosity and permeability samples must be extracted and dried. Remember, the porosity sample must be selected from as near as possible to where the saturation sample was selected for pore volume comparison.

After cutting and sizing the porosity and permeability samples, all fluids must be removed before running tests. Oil and water may be removed by either extraction in the Soxhlet apparatus or in an evaporating dish with a suitable solvent such as toluene or chlorothene. Upon completion of the extraction, the solvent is driven off by heating. Both porosity and permeability samples may be extracted and dried together, but it is critical that excessive heat not be used in drying; this would break down the hydratable clays and produce mineral alteration. Only the solvent and interstitial water should be removed in the drying process. The test samples should then be cooled and stored in a desiccator until ready for use. In the case of poorly consolidated samples, it is desirable to extract the fluids before final shaping.

23. In the case of low saturation with medium to high gravity oil, the samples may be extracted using the evaporating dish and solvent method. This is done by marking the samples for identification, placing them in an evaporating dish or pan, and heating the solvent to about 150°F. The container should have a tinfoil or regular cover. After heating for five to ten minutes, the container should be checked under the fluorescent light for color. The solvent should then be drained and a second application of solvent added to the samples and heated. This process should be repeated until all oil has been extracted from the samples and there is no fluorescence in the extraction solvent under the ultraviolet light.

24. If fluorescence examination indicates the sample is highly saturated or that the oil is medium to low gravity, the sample should be extracted by using the Soxhlet extractor as described in Paragraph 4.12. The glass flask is filled approximately two-thirds full of toluene, and the temperature-controlled hot plate is turned up to boil the solution. The extracting process continues as long as the hot plate is left on. The length of time that samples are left in the extractor will vary considerably. For adequate extraction, one run of solvent should be made on the samples that will yield no fluorescence of the solvent under the ultraviolet light.

25. Test samples should be dried in a drying oven with a vacuum fan. If this oven is not available, then the heat-lamp oven may be used. Tinfoil should be laid underneath the test samples. Drying the samples in the oven, whether they have been extracted with solvent or not, should take at least twenty minutes for a friable sand and up to forty minutes for a tight, silty sand. The sample should be cooled for at least ten minutes in a drying oven with the heat element off. After cooling, the porosity sample should be weighed in grams to the nearest 1/100 of a gram. This is the weight measurement for dry bulk density calculations and weight volume relationships on saturation pore volume. Upon completion of drying and cooling, the samples are ready for porosity measurement and permeability tests.

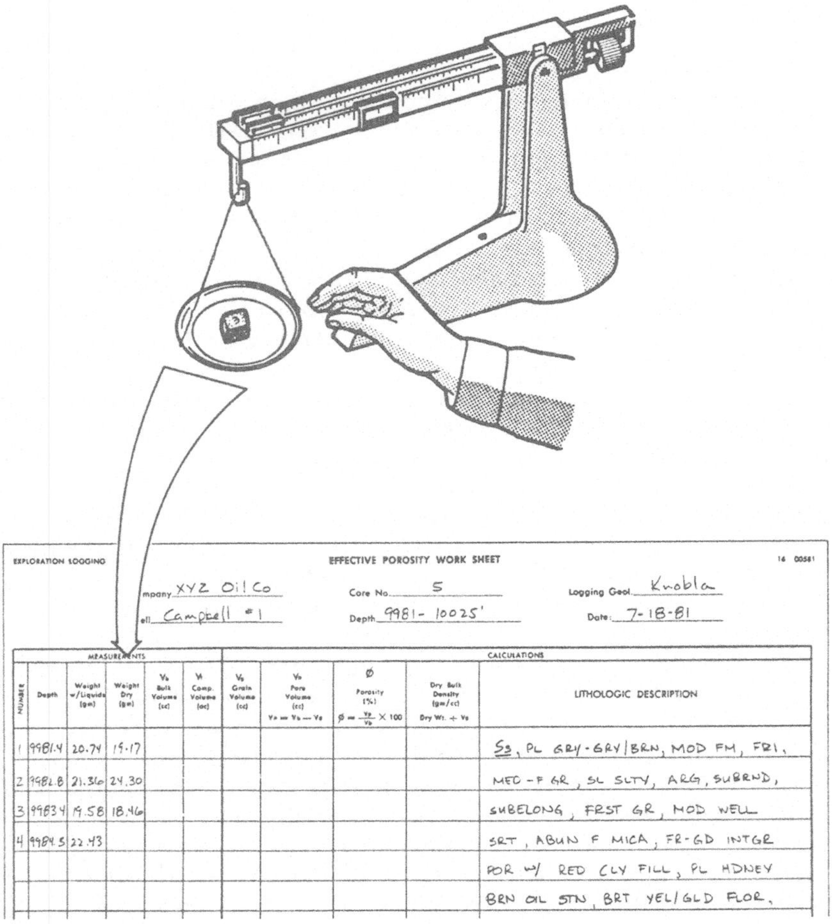

EXPLORATION LOGGING
EFFECTIVE POROSITY WORK SHEET
16 00581
mpany XYZ Oil Co
Core No. 5
Logging Geol. Knobla
ell Campbell #1
Depth 9981- 10025'
Date: 7-18-81
MEASUREMENTS
CALCULATIONS
NUMBER
Depth
Weight w/Liquids (gm)
Weight Dry (gm)
Vb Bulk Volume (cc)
Comp. Volume (cc)
Vg Grain Volume (cc)
Vp Pore Volume (cc) Vp = Vb — Vg
Ø Porosity (%) Ø = Vp/Vb × 100
Dry Bulk Density (gm/cc) Dry Wt. ÷ Vb
LITHOLOGIC DESCRIPTION
1 9981.4 20.74 19.17
2 9982.8 21.36 24.30
3 9983.4 19.58 18.46
4 9984.5 22.43
Ss, PL GRY-GRY/BRN, MOD FM, FRI,
MED-F GR, SL SLTY, ARG, SUBRND,
SUBELONG, FRST GR, MOD WELL
SRT, ABUN F MICA, FR-GD INTGR
POR w/ RED CLY FILL, PL HDNEY
BRN OIL STN, BRT YEL/GLD FLOR,

Figure 4-35. Weigh Porosity Sample and Record Dry Weight

26. Make note of the weight on the Effective Porosity worksheet (Figure 4-35). Once the samples are clean, dry and cool, and the porosity samples have been weighed, set them out in their correct order from left to right, making sure that all samples are clearly numbered (Figure 4-36).

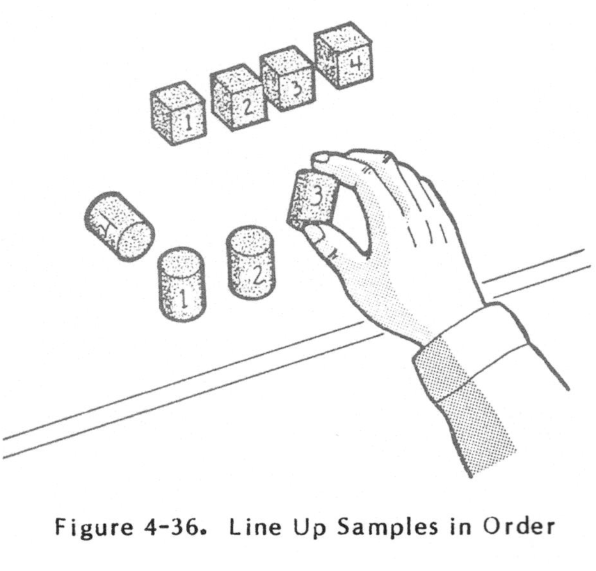

1
2
3
4
3
1
2

Figure 4-36. Line Up Samples in Order

27. The porosity samples can be set on the bench behind the mercury pump (porometer) and the permeability samples on the bench ready for mounting or on top of the permeameter, out of the way.

28. Having first checked that the pressure gauge reads zero, open the porometer lid and be sure that there is no mercury up in the lid (Figure 4-37). The sliding scale pointer should be about halfway (20 cc).

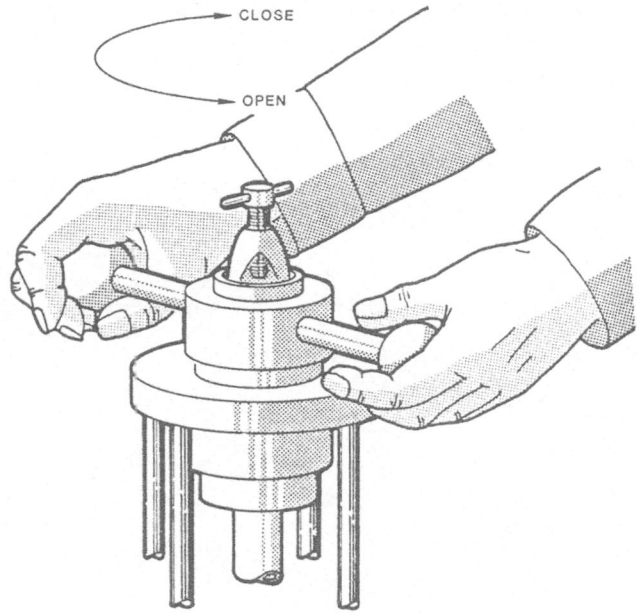

Figure 4-37. Open Porometer

29. Be sure the O-ring and seal are free of mercury, and insert porosity sample 1 (Figure 4-38). The top of the sample should be below the seat.

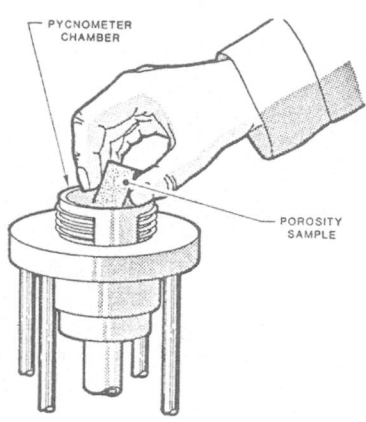

Figure 4-38. Insert Porosity Sample

30. Be sure the screw valve is open and the valve to the low-pressure gauge is open (counterclockwise), and close the porometer lid -- 1/6-turn clockwise -- and tighten as hard as you can.

31. Fill the pycnometer chamber with mercury by <u>slowly</u> turning the wheel until a small bead of mercury appears in the valve seat. If you overshoot, go back half a turn and come up again. All readings on the porometer must be taken with the wheel holding pressure, i.e., going up clockwise to the reading (Figure 4-39).

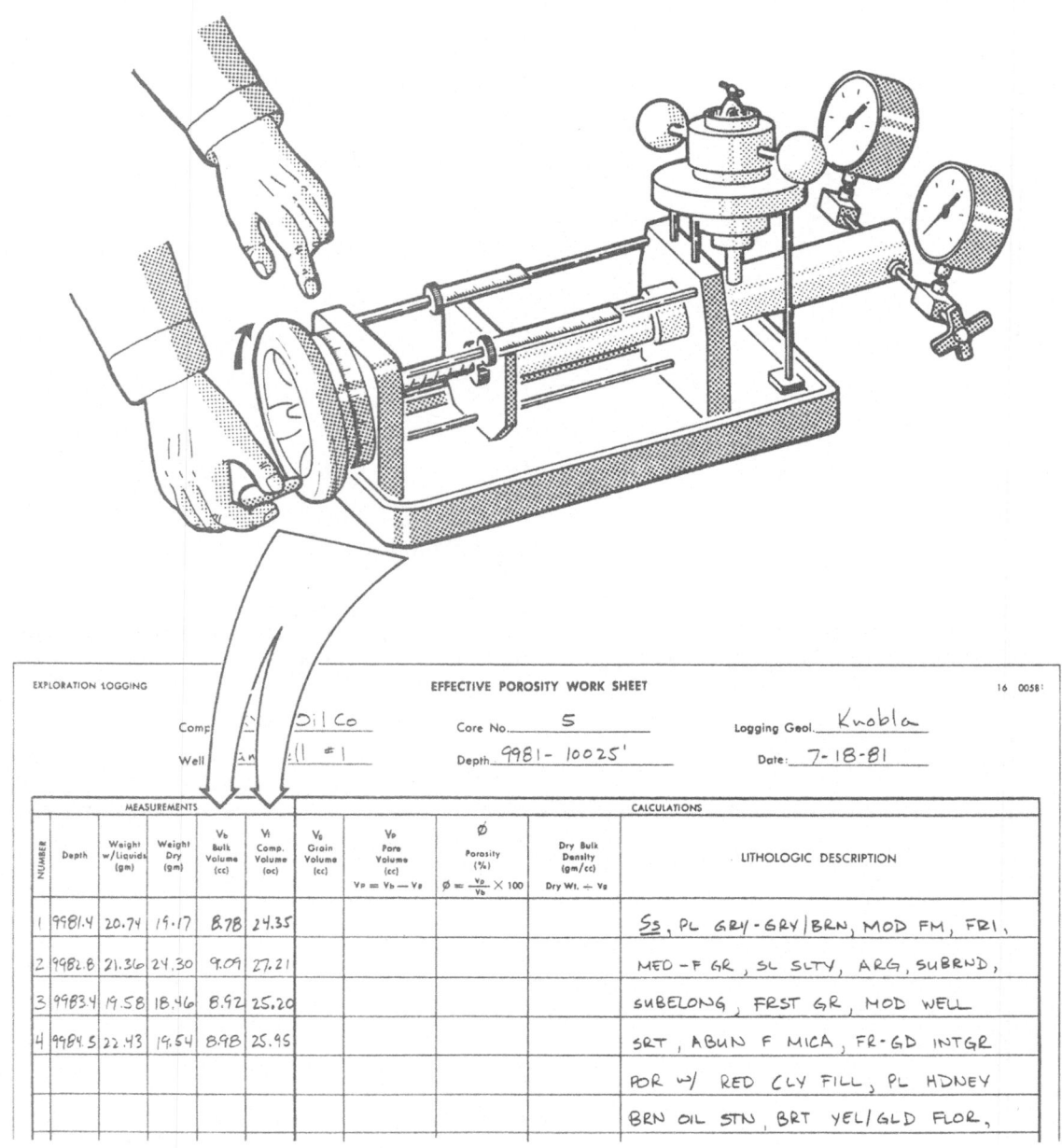

EXPLORATION LOGGING					EFFECTIVE POROSITY WORK SHEET				16 0058

Comp: Oil Co   Core No. 5   Logging Geol. Knobla

Well: well #1   Depth 9981- 10025'   Date: 7-18-81

NUMBER	Depth	Weight w/Liquids (gm)	Weight Dry (gm)	Vb Bulk Volume (cc)	Vi Comp. Volume (cc)	Vg Grain Volume (cc)	Vp Pore Volume (cc) Vp = Vb — Vg	Ø Porosity (%) Ø = Vp/Vb × 100	Dry Bulk Density (gm/cc) Dry Wt. ÷ Vg	LITHOLOGIC DESCRIPTION
1	9981.4	20.74	19.17	8.78	24.35					<u>Ss</u>, PL GRY-GRY/BRN, MOD FM, FRI,
2	9982.8	21.36	24.30	9.09	27.21					MED -F GR, SL SLTY, ARG, SUBRND,
3	9983.4	19.58	18.46	8.92	25.20					SUBELONG, FRST GR, MOD WELL
4	9984.5	22.43	19.54	8.98	25.95					SRT, ABUN F MICA, FR-GD INTGR
										POR W/ RED CLY FILL, PL HDNEY
										BRN OIL STN, BRT YEL/GLD FLOR,

Figure 4-39. Record Bulk Volume

32. Record the bulk volume measurement by reading off the cc's on the upper sliding scale, and decimals on the inner circular scale. The reading should be noted on the Effective Porosity worksheet (Figure 4-39).

33. Run the mercury back by turning the wheel counterclockwise to the ink mark at 42+ on the sliding scale, making sure that you don't go back too far. Come up to the 40 cc mark on the upper sliding scale by rotating clockwise, and exactly to zero on the inner circular scale. If you overshoot, go back one-half turn and come up again.

34. Firmly tighten down the needle valve, using finger tip pressure only, being careful not to overtighten as this may damage the thread and/or the seat (Figure 4-40).

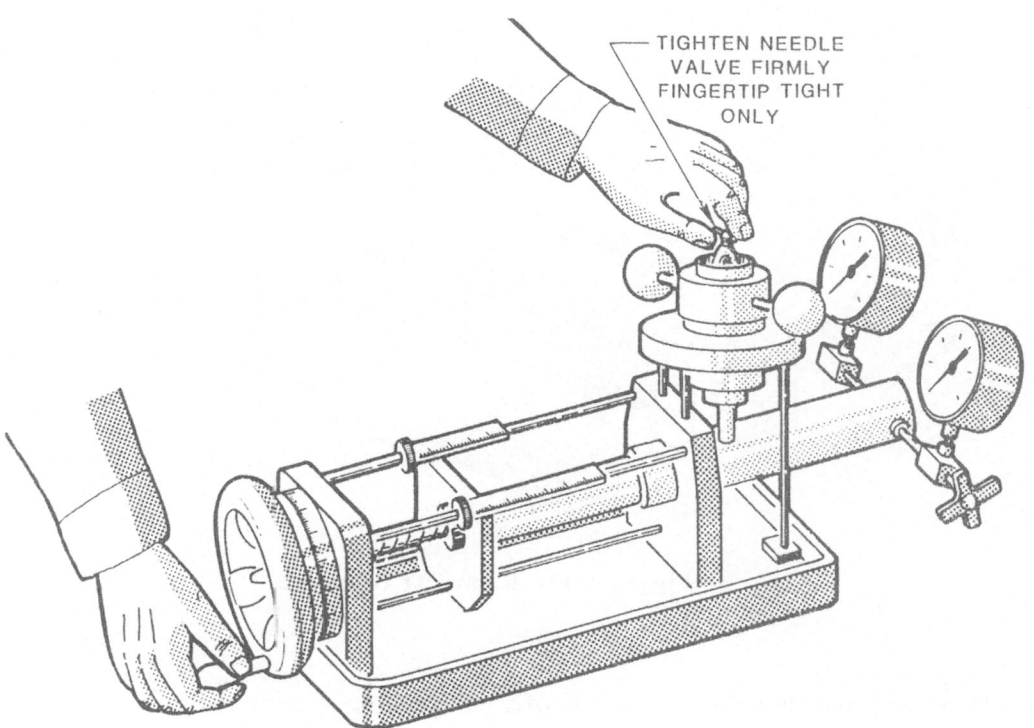

TIGHTEN NEEDLE VALVE FIRMLY FINGERTIP TIGHT ONLY

Figure 4-40. Close Needle Valve Finger Tight

35. Bring pressure up to 30 psi by rotating the wheel clockwise.

36. Record the compression volume by reading the cc's on the sliding scale and the decimals on the inner circular scale. The readings should be recorded on the Effective Porosity worksheet (Figure 4-39).

37. Turn the wheel back until the pressure drops to zero, making sure that you do not turn back beyond the 40 cc mark (Figure 4-41).

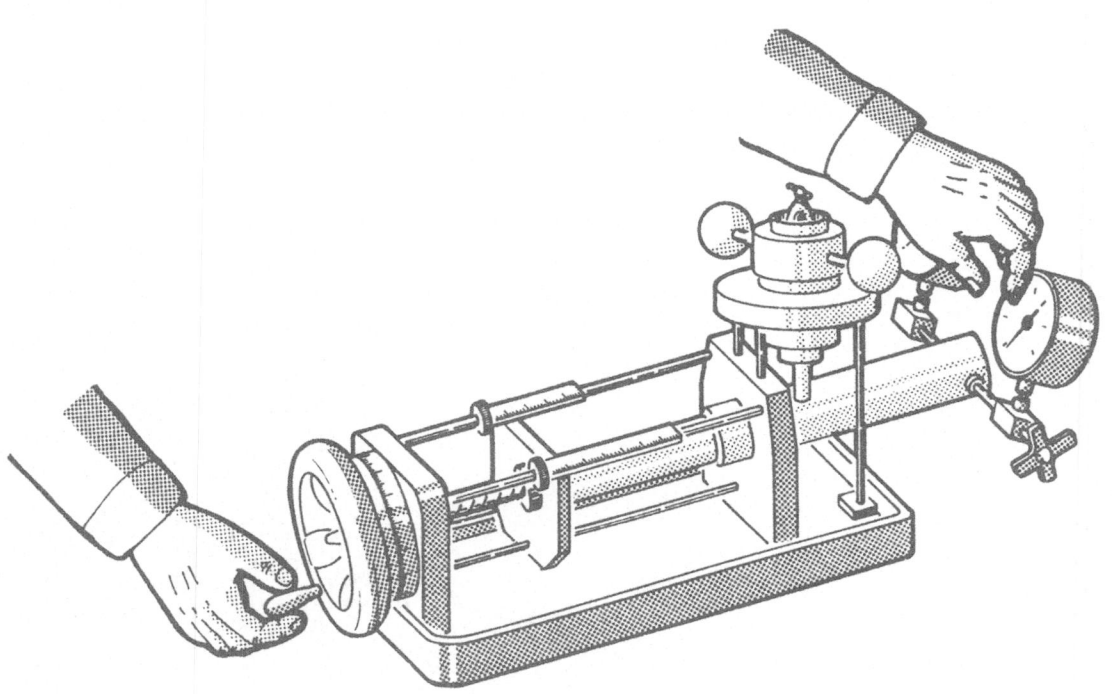

Figure 4-41. Back-Off Pressure

38. Double-check that the pressure is off, and then open the needle valve.

39. Carefully open the lid and lift it off carefully to avoid spilling mercury. Then remove the sample and put it back in order in its tray.

40. Clean the pump by brushing off the O-ring and seal, and clean off excessive dust and sand grains from the mercury (Figure 4-42).

41. Repeat the operation for the next sample until all are done.

WARNING

Keep the O-ring and seal free of mercury and dirt. Always wind up to reading, i.e., turn the handle clockwise, and keep the slack taken up. Take care not to draw air into the system, and avoid spilling mercury. Mercury is poisonous and expensive.

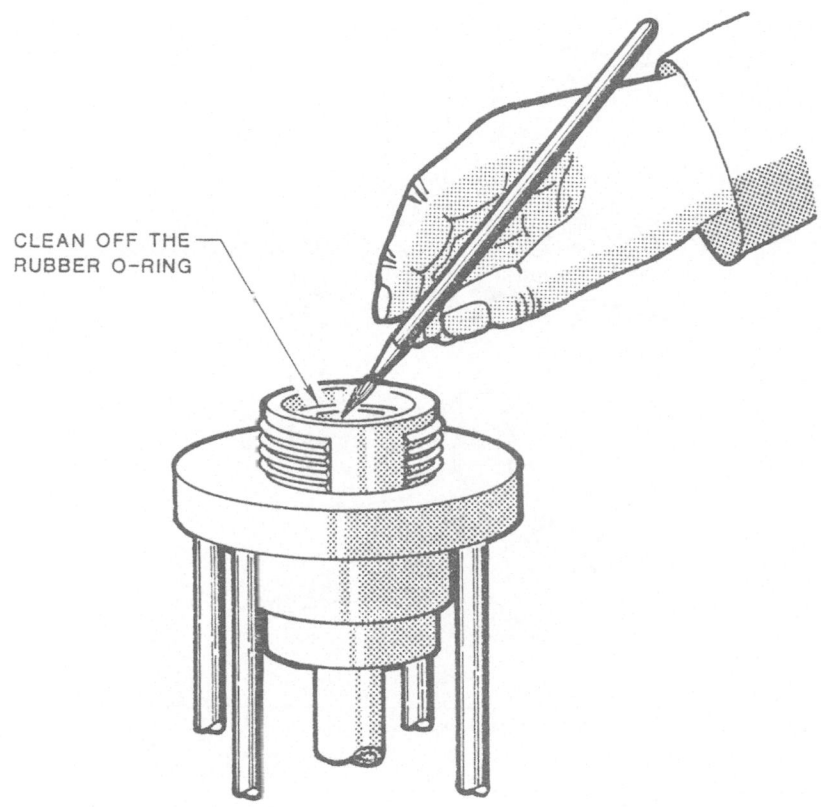

CLEAN OFF THE
RUBBER O-RING

Figure 4-42. Keep O-Ring, Seal and Mercury Surface Clean

NOTE

The saturation samples should now be ready for you
to take the oil reading.

42. Remove the sample 1 collecting tube (left-hand first; Figure 4-43).

43. Take the oil reading, ensuring a good oil/water separation by centrifuging or rubbing between the hands. Take an average meniscus reading. Make a note of the reading on the Saturation worksheet and also make a note of the color and any other obvious property; e.g. paraffinic.

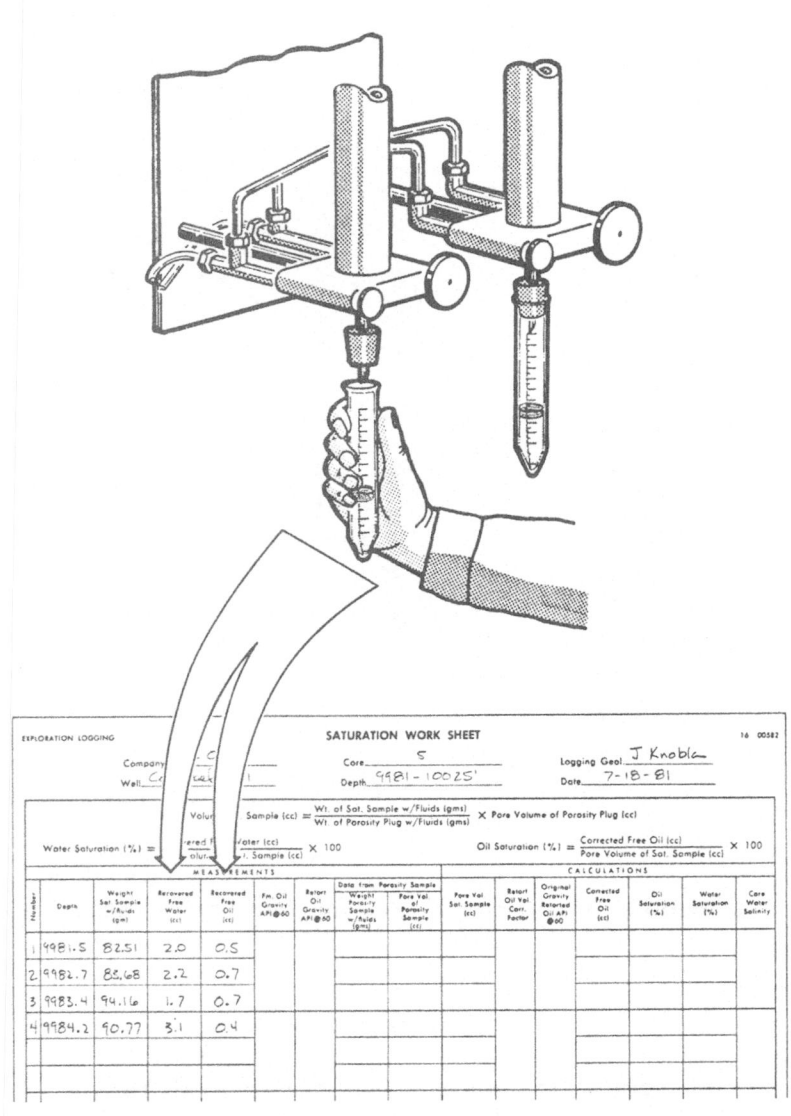

Figure 4-43. Remove Collecting Tubes and Record Recovered Water and Oil

44. Repeat Steps 42 through 43 for sample 2.

45. Separate or pipette off the oil from the tubes and keep them for oil gravity tests (Figure 4-44).

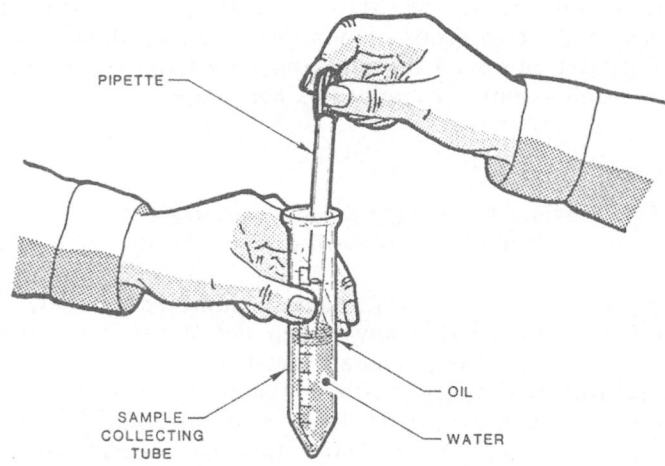

Figure 4-44. Separate Oil for Gravity Test

46. Remove the retort bombs from the stills and put them in the sink or in a metal bucket. <u>Use gloves</u> and do not wet the hot bombs; allow them to cool in air (Figure 4-45).

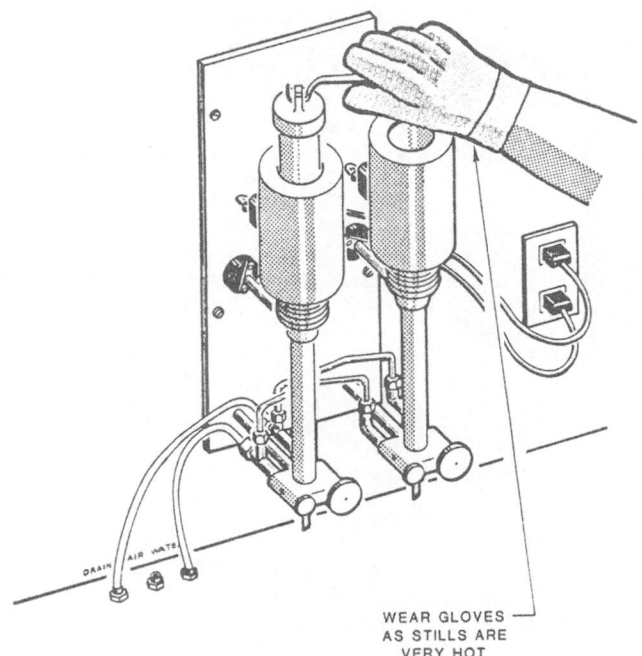

Figure 4-45. Remove Retorts

47. Clean up, wash and dry the collecting tube, and run a cleaning brush through the stills (wear gloves).

48. Turn the water on and ensure that it is circulating. Put two fresh collecting tubes on the stills, each with two drops of demulsifier, and place retort bombs 3 and 4 in the stills, having first removed the masking tape from their tips. Set the timer and repeat the saturation sample procedure as for 1 and 2.

NOTE

Steps 49 through 51 may be omitted if
the plugs are well consolidated.

49. Line up the dry, clean permeability plugs in numerical order, making sure that all measurements (area and length) have been noted for each plug. Line up all the cylindrical plugs on top of the permeameter (Figure 4-46). Set out as many sleeves as you have noncylindrical plugs on a board or piece of asbestos sheet. Pour a small amount of sand into each sleeve to form a small conical heap, and put the plugs into the sleeves -- pressing them lightly into the sand. Keep them in numerical order from left to right (Figure 4-47).

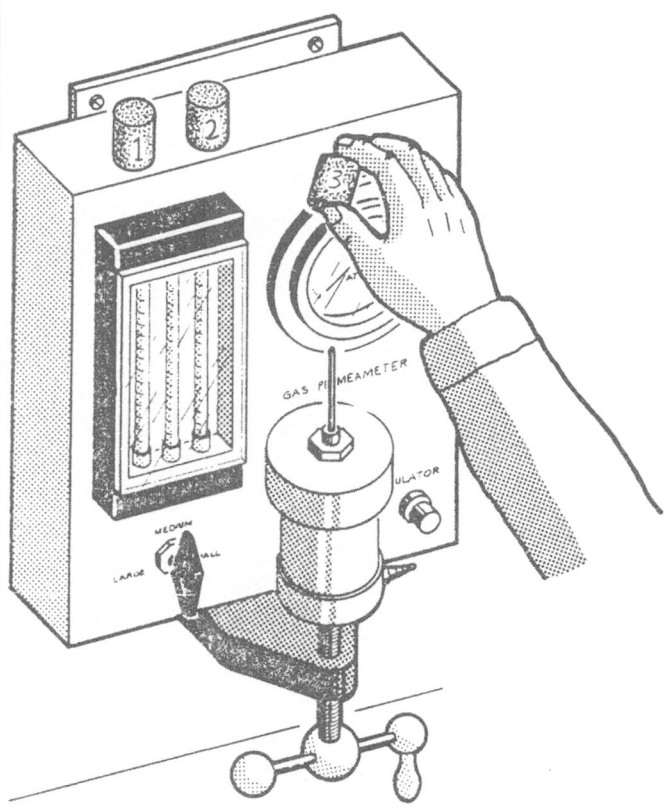

Figure 4-46. Line Up Permeability Plugs in Order

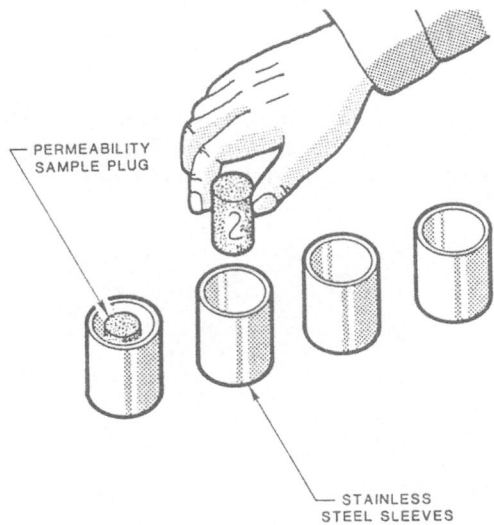

Figure 4-47. Insert Plugs into Sleeves

NOTE

Retorts should now be ready
for water readings.

50. Heat the sealing wax on a hot plate, using the enameled sauce pan. When liquid, pour the wax carefully into the sleeves, filling the annulus between the sleeves and the samples. The sealing wax should just reach the edge of the plug without covering the end, as shown in Figure 4-48. Avoid getting wax on the end of the sample; if you do, it will have to be scraped off with a knife. Make certain that the seal between the plug and the sleeve is complete. It is best to arrange to seal all the test samples which require wax sealing at one time, if possible.

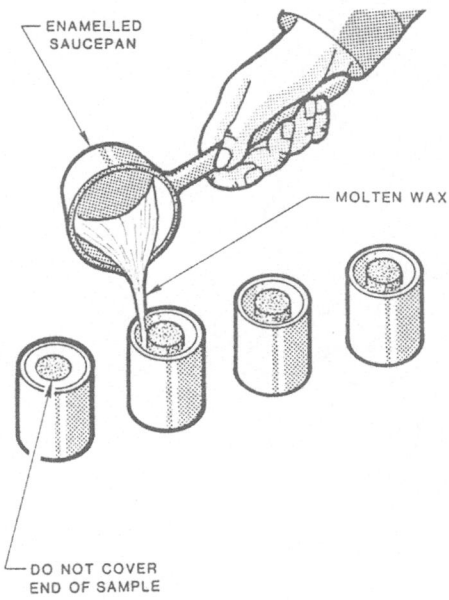

Figure 4-48. Pour Wax Into Sleeves

51. When the wax has cooled and hardened, mark the appropriate sample number on the wax at the end of the sleeve as shown in Figure 4-49. <u>Do not</u> place it in the permeameter until the wax has completely hardened.

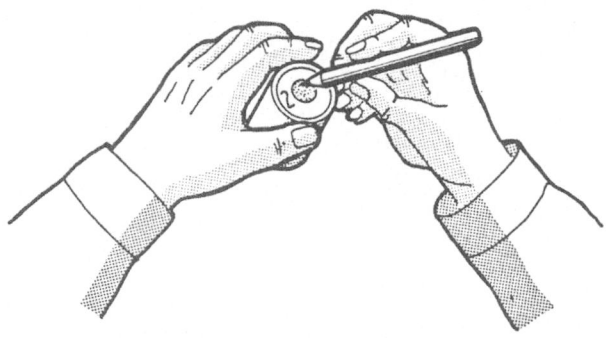

Figure 4-49. Mark Sample Number on Wax

52. Set plug number 1 into the permeameter core holder. If the plug is cylindrical, push it into the large rubber-core sleeve or, if it is mounted, into the tubular rubber sleeve. Press the rubber sleeve complete with sample into the metal core holder, tapered-end downward as shown in Figure 4-50.

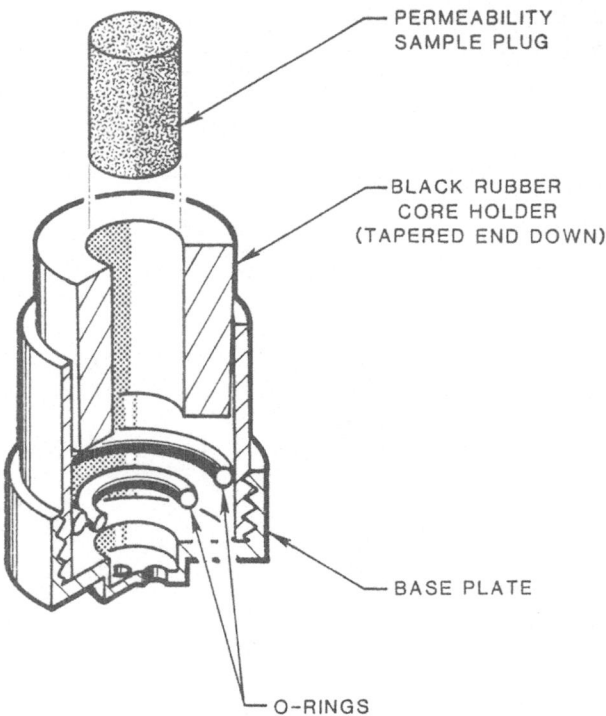

PERMEABILITY
SAMPLE PLUG

BLACK RUBBER
CORE HOLDER
(TAPERED END DOWN)

BASE PLATE

O-RINGS

Figure 4-50. Insert Sample in Core Holder

53. Screw on the bottom of the core holder, making sure that the rubber O-rings and/or metal washer are correctly positioned. Hand-tighten only with fingertip pressure (Figure 4-51).

54. Mount the core holder in the screw mounting in the permeameter, making sure the top of the rubber sample sleeve is against the upper metal fitting. Also, ensure that the metal core holder does not catch on the lower portion of the upper metal fitting. Tighten-up hard on the handle to ensure a good fit around the sides of the sample. Also, make sure that the rubber sleeve is not blocking the air outlet in the metal base (Figure 4-51).

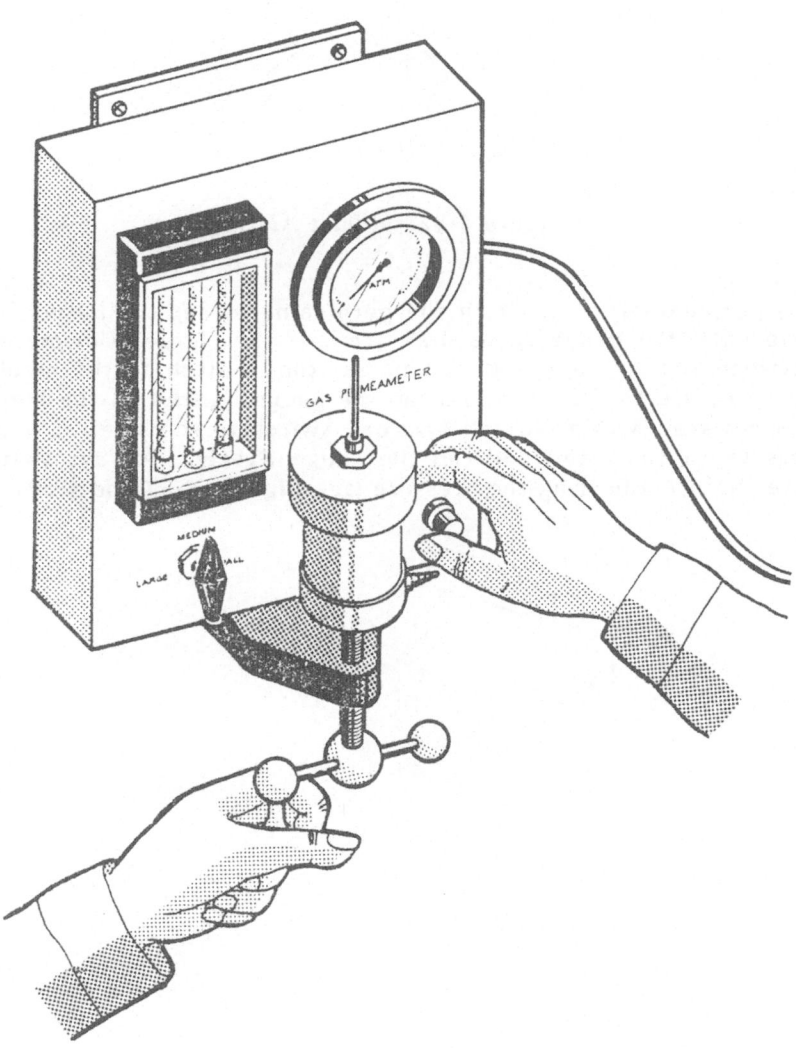

Figure 4-51. Mount Core Holder and Tighten

144

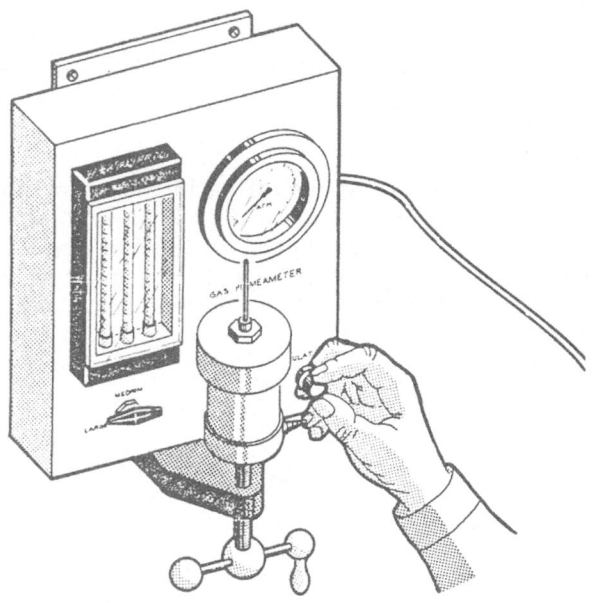

Figure 4-52. Adjust Airflow

55. Turn the permeameter on. With the permeameter set to "large" flowmeter, open the valve until the gauge reads 0.25 atm. If the "large" flowmeter ball does not stay between the 2.0 and 14.0 cm marks, then switch to the medium flowmeter, raise the pressure to 0.50 atm and tap the gauge lightly. If the medium flowmeter ball does not stay within 2.0 and 14.0 cm, switch to the small flowmeter, and raise the pressure to 1.00 atm, and tap the gauge lightly. Always switch to the next flowmeter before adjusting the pressure (see Figures 4-52 and 4-53).

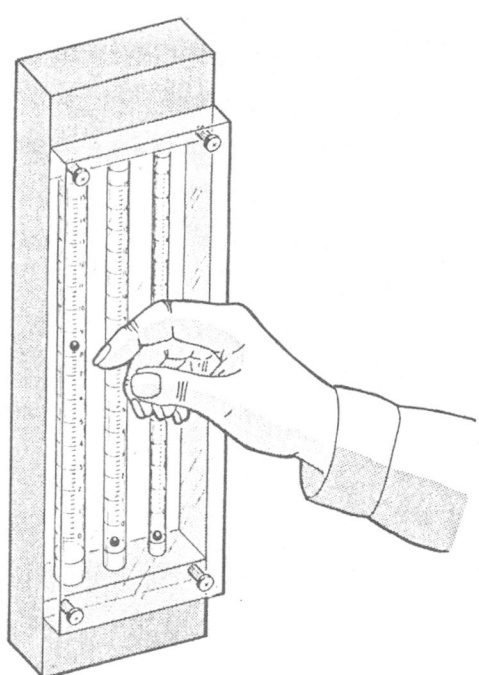

Figure 4-53. Read Flowmeter

56. Make notes of the flowmeter reading, flowmeter size, inlet pressure, and air temperature on the Permeability worksheet (Figure 4-54).

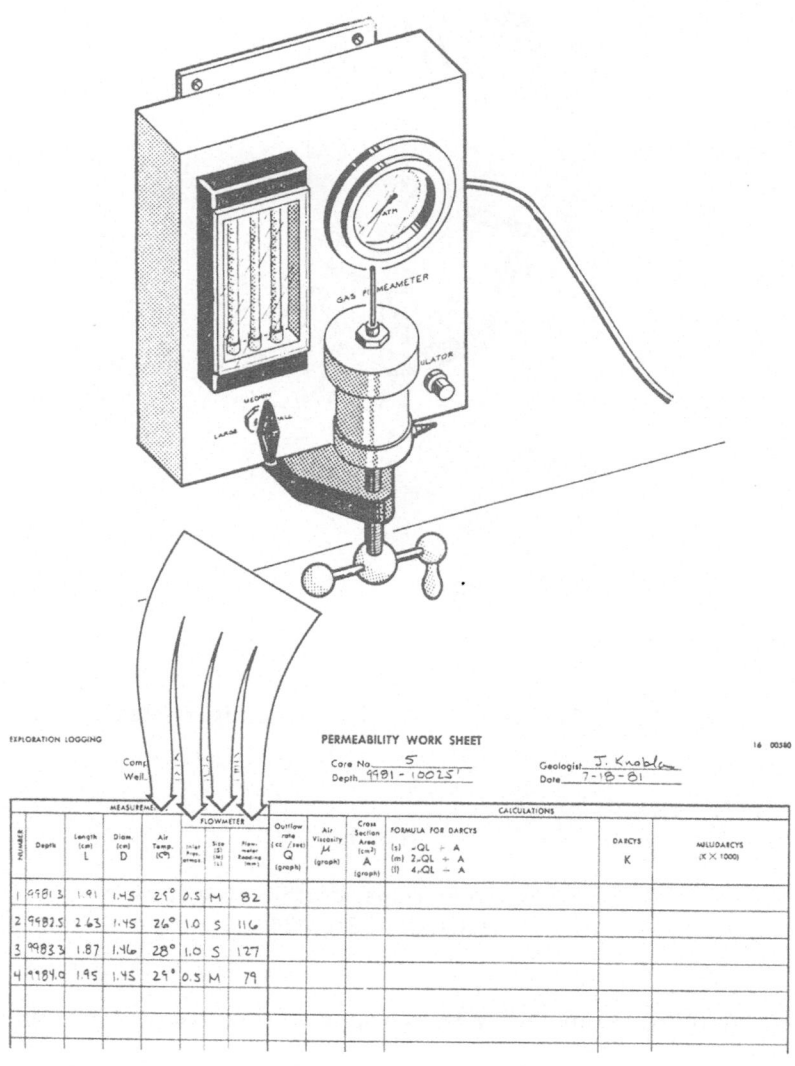

PERMEABILITY WORK SHEET

Core No. ____5____   Geologist _J. Knoblan_
Depth _9981 - 10025'_   Date _7-18-81_

16   00380

NUMBER	Depth	Length (cm) L	Diam. (cm) D	Air Temp. (C°)	Inlet Press. (psig)	Size (s) (m) (L)	Flow-meter Reading (mm)	Outflow rate Q (cc/sec) (graph)	Air Viscosity μ (graph)	Cross Section Area (cm²) A (graph)	(s) QL ÷ A (m) 2QL ÷ A (L) 4QL ÷ A	DARCYS K	MILLIDARCYS (K × 1000)
1	9981.3	1.91	1.45	25°	0.5	M	82						
2	9982.5	2.63	1.45	26°	1.0	S	116						
3	9983.3	1.87	1.46	28°	1.0	S	127						
4	9984.0	1.95	1.45	29°	0.5	M	79						

Figure 4-54.   Record Permeameter Readings

57. Turn the pressure off, take out the holder, remove the rubber sleeve, and remove the sample plug. Keep the permeameter and porosity plugs for the client or for rerunning (Figure 4-55).

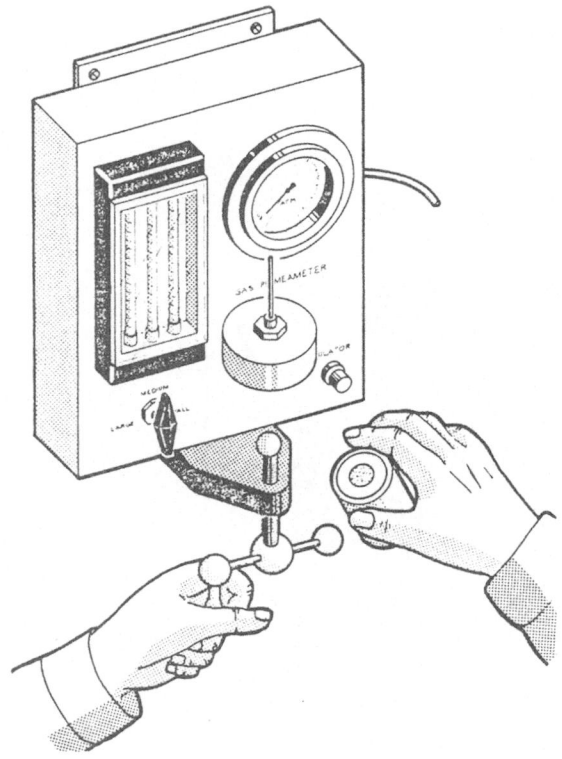

Figure 4-55. Remove Core Holder and Insert Next Sample

NOTE

When the timer rings, drain water from
stills as in Step 21.

58. Run the remaining permeability samples, repeating Steps 51 through 56.

NOTE

Retorts will probably now be ready
for final oil reading.

59. Take the oil readings as for the previous samples and load up the next two.

60. When cooled, the used retorts have to be cleaned out. <u>Wearing gloves,</u> remove the retorts from the sink or bucket and remove the tops (use vise and bar, if necessary). Throw out the used core and clean the retorts with toluene or acetone by washing through, using two beakers as illustrated in Figure 4-56. Make sure that the retort spouts are clean.

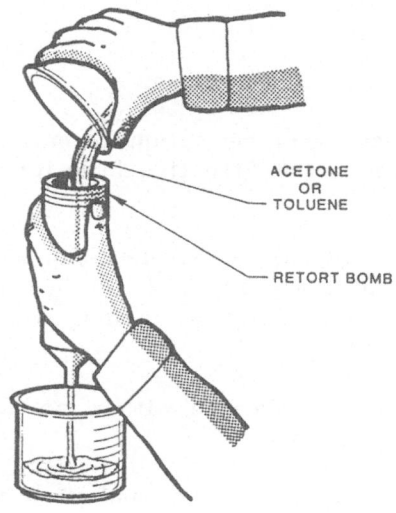

ACETONE
OR
TOLUENE

RETORT BOMB

Figure 4-56. Clean Retorts when Cool

61. Clean out the collecting tubes by washing with soap and water, using a test-tube brush. They should then be dried and put away thoroughly clean for the next use (Figure 4-57).

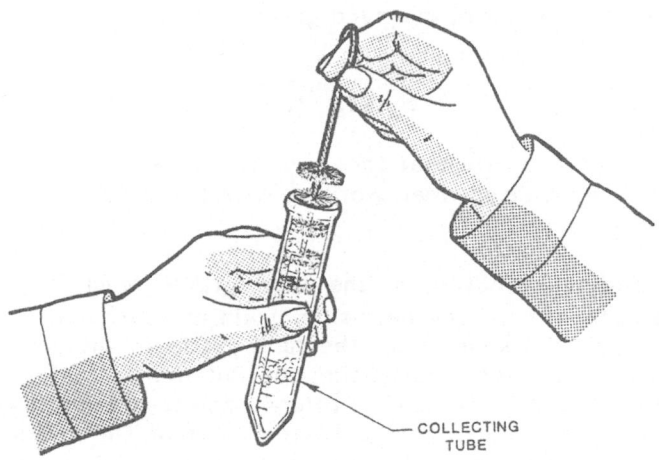

COLLECTING
TUBE

Figure 4-57. Clean Collecting Tubes

62. Run any remaining saturation, porosity and permeability samples as described in Steps 1 through 61.

63. Repeat any tests which seem doubtful, even if it is necessary to break out more core.

64. Put all remaining samples and plugs into labeled sample sacks and deliver them to the client's representative.

4.29 Optional Total Porosity Test

If total porosity is required, one extra measurement must be made in addition to the measurements already recorded on the Effective Porosity Worksheet. This involves a glass volumeter.

1. Obtain the bulk volume of toluene in the glass volumeter whereby the liquid level reaches the bottom part of the calibrated tube when the volumeter is inverted. This will provide a base reference volume reading before adding sand grains to the volumeter. Record this reading.

2. Crush the porosity sample to grain size with a mortar and pestle. Be <u>extremely</u> careful not to lose any sand grains.

3. Place the volumeter in the upright position and allow all the toluene to drain from ther upper part of the volumeter. Remove the cap and carefully transfer the weighed sand grains to the volumeter cap. Replace the cap carefully so as not to lose any sand grains, and then invert the volumeter for the final reading. Be sure that all sand grains stay in the cap of the volumeter, that the grains are completely wetted by toluene, and that all entrapped air is eliminated from the volume of sand grains and toluene. Record this final reading.

4. Initial (R_1) and final (R_2) readings are taken from the volumeter before and after the sand grains are added. The increase in volume of fluid as a result of adding the sand is a direct measurement of the sand grain volume.

WARNING

Make sure that the work area is well
ventilated when working with toluene.

5. After the R_2 reading is obtained, the sand grains must be removed from the volumeter cap to prepare for the next sand grain measurement. Tilt the volumeter at a maximum angle, still keeping all the sand grains in the cap. Place over a sink and remove the cap, promptly uprighting the bottom part of the volumeter. The objective is to lose the least amount of toluene and to prevent any sand grains from being transferred from the cap to the bottom part of the volumeter. Remove all sand grains from the cap and clean out any residue. Add sufficient toluene to replace that lost, whereby the liquid level will again provide a reference volume reading in the inverted position. The volumeter is then ready for the next test.

6. Record the results on the Total Porosity Worksheet (Figure 4-58).

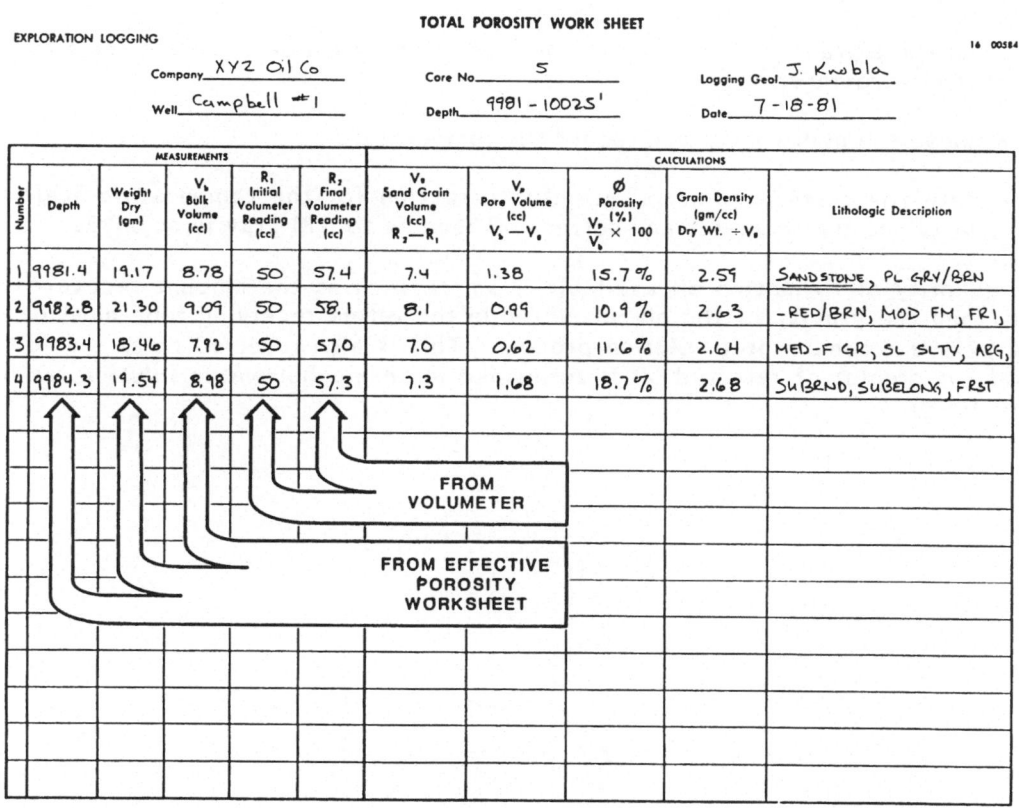

Figure 4-58. Total Porosity Worksheet

4.30 Oil Gravity Tests

The gravity of the oil is necessary for the proper correction of recovered oil volume on the saturation tests. It is also an important factor in evaluating all core analysis data. Whenever possible, the gravity of the oil should always be reported on the core analysis report. Oil gravities are corrected to API gravity at 60°F. If the specific gravity of the oil is required and the API gravity is known, it can be determined from the table that follows Figure 4-59.

The equipment consists of two API Oil Gravity Hydrometers and a clear glass cylinder for suspending the hydrometers in the oil or 'alcohol-water' solution. One hydrometer range is from 10° to 45° and the other is from 45° to 90°. Both hydrometers are equipped with thermometers and temperature conversion scales for converting to API gravity at 60°F.

When sufficient sample of formation oil is available to float the hydrometer:

1. Fill glass cylinder one-half full with oil sample and place appropriate hydrometer in cylinder. Be sure the hydrometer floats freely and does not stick to the walls of the cylinder or touch the bottom of the cylinder.

2. Record the gravity. This will be indicated by the level of oil on the scale at the top of the hydrometer.

3. Remove the hydrometer and record the temperature.

4. Locate the corresponding oil gravity correction for the temperature scale to the right of the thermometer and correct oil reading to API gravity at 60°F.

If the quantity of formation oil available is small, such as the amount recovered in the retort, then the gravity of the oil recovered in the retort receiving tube must be determined by the alcohol-water solution process. This is also referred to as the 'oil drop' method. A droplet of retorted oil is suspended in an alcohol-water solution, as follows (Figure 4-59):

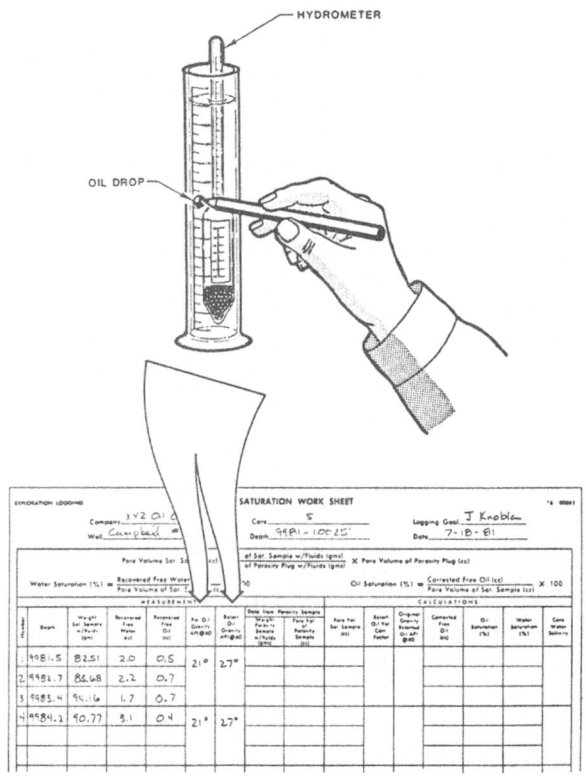

Figure 4-59. Oil Drop Gravity Test

1. Fill a glass cylinder half-full with an alcohol-water solution of approximately 1 to 1.

2. Pipette 1 drop of oil from the fluid saturation recovery into the solution.

3. Stir solution gently and note whether the oil drop rises or falls. Progressively add water to make gravity of the solution lower, or add alcohol to make gravity higher, depending on whether the drop of oil falls or rises. Repeat until drop remains suspended at a constant level after gentle stirring of the solution.

4. Measure the gravity of the alcohol-water solution with an API hydrometer.

The conversion factors listed below may be used for converting API gravity to specific gravity and for determining the density of oil in pounds per cubic foot.

^{o}API	Specific Gravity	lb/ft^3
8	1.014	63.4
10	1.000	62.5
12	0.986	61.6
14	0.973	60.8
16	0.959	60.0
18	0.946	59.1
20	0.934	58.4
22	0.922	57.6
24	0.910	56.9
26	0.898	56.1
28	0.887	55.4
30	0.876	54.7
32	0.865	54.1
34	0.855	53.4
36	0.845	52.8
38	0.835	52.2
40	0.825	51.6
42	0.815	51.0
44	0.806	50.4
46	0.797	49.8
48	0.788	49.3
50	0.780	48.7
52	0.771	48.2
54	0.763	47.7
56	0.755	47.2
58	0.747	46.7
60	0.739	46.2

$$^{o}API @ 60°F = \frac{141.5}{G} - 131.5$$

G = Specific Gravity @ 60°F

4.31 Water Salinity Tests

There are two methods for determining core-water salinity: chemical titration and resistivity measurements. Both methods measure the amount of salts extracted from a weighed volume of sample. These measured values must be converted to milligrams of salt leached from the sample and then calculated to parts-per-million total salt, the normal method for expressing salinity.

The following equipment is required for core-water salinity determination:

- pipettes and chemicals for titration
- a mud resistivity meter and thermometer for resistivity measurements
- a mortar and pestle for crushing the sample
- a triple-beam balance for weighing the sample, an evaporating dish
- a temperature-controlled hotplate

Twenty grams of sample are required for core-water salinity determination. Preferably, the sample should be a portion of the retorted sample used in the saturation test. If not, it should be selected at some point in the core as close as possible to the saturation or porosity test sample. If the sample has not been previously retorted it must be dried in the sample drying oven before the salinity test is run.

1. Titration Method: Grind to grain size approximately 20 to 25 g of sample from a previously retorted saturation sample.

2. Weigh 20.0 g of sample and transfer to an evaporating dish. Add 100 cc of distilled water.

3. Heat the solution on the temperature-controlled plate. <u>Do not boil.</u> Stir the solution often.

4. Filter the solution.

5. Run a salinity titration of 1.0 cc of the leached sample. The result should be expressed as parts of sodium chloride per million parts of core water. This is an intermediate measurement and not the final calculation of core-water salinity.

6. The measurement from the titration method is calculated in milligrams of NaCl, leached from the sample by the formula:

$$NaCl\ (mg) = \frac{cc\ AgNo_3 \times 1.65 \times vol\ H_2O\ used\ to\ leach\ sample\ (cc)}{cc\ of\ filtrate\ used}$$

1.65 is the factor for converting mg chloride ion to mg NaCl while the silver nitrate solution is .0282N equivalent to 1 mg of Cl ion per ml.

7. The volume of the core water in the salinity sample is calculated as follows:

$$V_b = \frac{W}{\rho_m} \div (1 - \phi) \tag{4-11}$$

$$\text{Pore Water (ml)} = V_b \times \phi \times S_w \tag{4-12}$$

where

W   =   dry weight of salinity sample

ρ_m =   grain density of porosity sample

ϕ =   porosity from porosity test sample

V_b =   bulk volume of salinity sample

Sw   =   water saturation from saturation test

The dry bulk density may be used for the grain density of the salinity sample.

The salinity of the pore water is obtained by:

$$\frac{\text{(mg) NaCl}}{\text{(ml) Pore Water}} \times 100 = \tag{4-13}$$

mg NaCl per 100 g pore water

mg NaCl per 100 g pore water x 10 = ppm NaCl   (4-14)

9. Resistivity Measurement Method: Measure resistivity of filtered solution with a Baroid or Fann Resistivity Meter.

10. Correct resistivity to 77°F by means of the Temperature Correction Chart provided with the resistivity meter. This result must also be corrected for pore water volume.

11. The resistivity value for the salt solution can be converted directly to milligrams of salt or salinity by using the conversion chart for salinity determination supplied with the resistivity meter.

4.32 CORE ANALYSIS CALCULATIONS

4.33 Effective Porosity

1. Using the previously established porometer calibration graph (see Paragraph 4.24), convert the Compression Volume (V_f) readings to Grain Volume (V_g) and record on the worksheet (Figure 4-60 and 4-61).

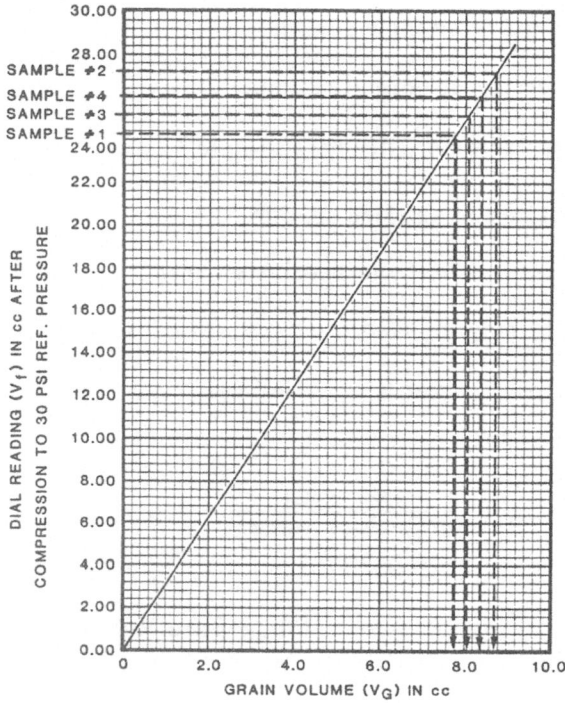

Figure 4-60. Porometer Calibration

2. Calculate Pore Volume (V_p) and record on worksheet:

$$V_p = V_b - V_g \qquad (4-15)$$

3. Calculate Porosity (ϕ) and record on worksheet:

$$\phi\% = \frac{V_p}{V_b} \cdot 100\% \qquad (4-16)$$

4. Calculate Apparent Dry Bulk Density and record on worksheet:

$$\rho_b = \frac{Dry\ Weight\ (gms)}{V_g\ (cc)} \qquad (4-17)$$

5. Type a final copy of the Effective Porosity Worksheet and brief lithology description abstracted from the Core Report (Figure 4-61).

EXPLORATION LOGGING   EFFECTIVE POROSITY WORK SHEET   16 00581

Company XYZ OIL CO   Core No. 5   Logging Geol. J KNOBLA

Well CAMPBELL #1   Depth 9981-10025'   Date: 7-18-81

NUMBER	Depth	Weight w/Liquids (gm)	Weight Dry (gm)	V_b Bulk Volume (cc)	V_l Comp. Volume (cc)	V_s Grain Volume (cc)	V_p Pore Volume (cc) $V_p = V_b - V_s$	ϕ Porosity (%) $\phi = \frac{V_p}{V_b} \times 100$	Dry Bulk Density (gm/cc) Dry Wt. ÷ V_s	LITHOLOGIC DESCRIPTION
1	81.4	20.74	19.17	8.78	24.35	7.70	1.08	12.3%	2.49	Ss, PL GRY-GRY/BRN, MOD FM, FRI, MED-F GR,
2	82.8	21.36	24.30	9.09	27.21	8.95	0.14	1.5%	2.38	SL SLTY, ARG, SUBRND, SUBELONG, FRST GR,
3	83.4	19.58	18.46	8.92	25.20	8.02	0.90	10.1%	2.30	MOD WELL SRT, ABUN F MICA, FR-GD INTGR
4	84.3	22.43	19.54	8.98	25.95	8.24	0.74	8.2%	2.37	POR W/ RED CLY FILL, PL HONEY BRN OIL STN,

Figure 4-61. Effective Porosity Worksheet

4.34 Permeability

1. Using the Permeameter Outflow Graph (Figure 4-62), convert the reading for the appropriate flowmeter to Outflow Rate (Q) and record on the Permeability Worksheet (Figure 4-63).

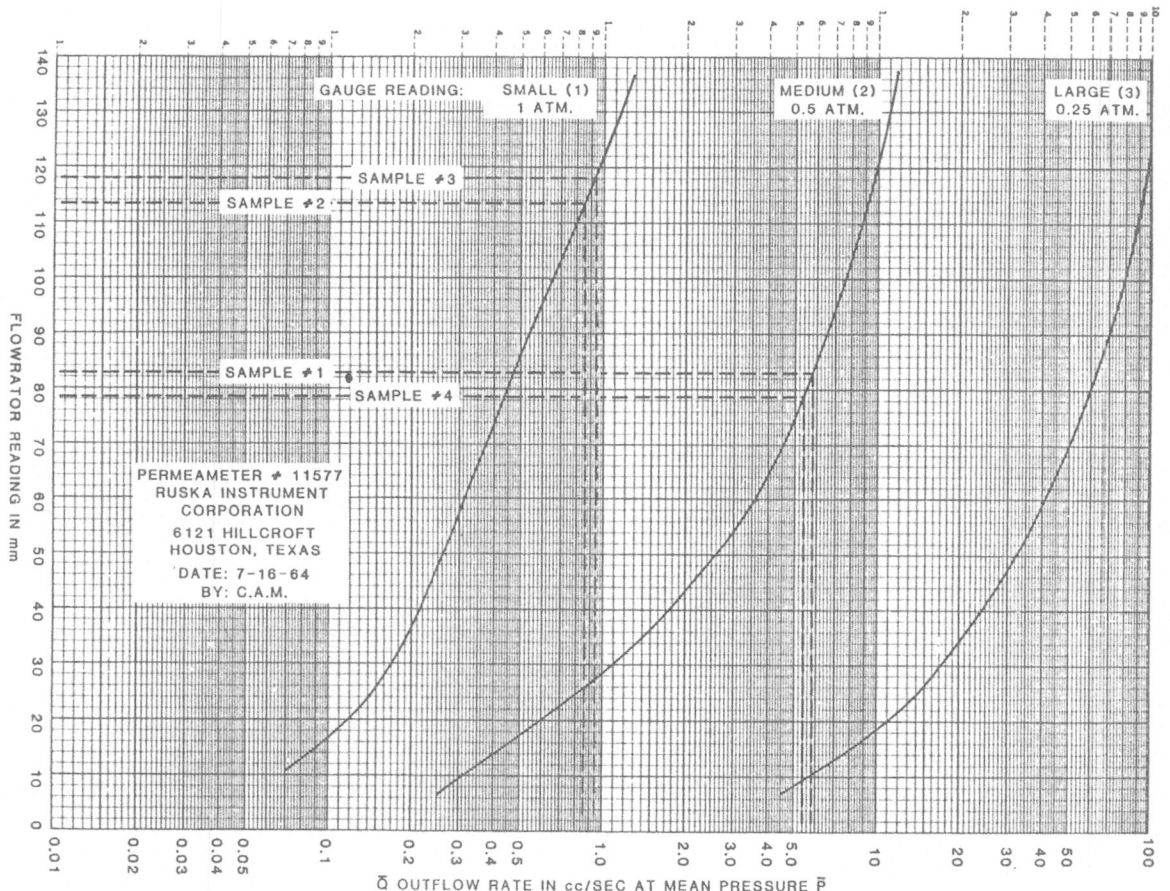

Figure 4-62. Permeameter Outflow Graph (Example Only)

PERMEABILITY WORK SHEET 16 00380

Company __XYZ Oil Co__   Core No. __5__   Geologist __J Knobla__
Well __Campbell #1__   Depth __9981-10025'__   Date __7-18-81__

NUMBER	Depth 9981-	MEASUREMENTS							CALCULATIONS						
		Length (cm) L	Diam. (cm) D	Air Temp. (C°)	FLOWMETER			Outflow rate (cc/sec) Q (graph)	Air Viscosity μ (graph)	Cross Section Area (cm²) A (graph)	FORMULA FOR DARCYS (s) μQL ÷ A (m) 2μQL ÷ A (l) 4μQL ÷ A	DARCYS K	MILLIDARCYS (K × 1000)
					Inlet Pres. (atmos.)	Size (S) (M) (L)	Flow-meter Reading (mm)						
1	81.3	1.91	1.45	29°	0.5	M	82	5.8	.0184	1.65	2(0.0184 * 5.8 * 1.91)/1.65	0.247	247
2	82.5	2.63	1.45	26°	1.0	S	116	0.80	.0182	1.65	(0.0182 * 0.8 * 2.63)/1.65	0.023	23
3	83.3	1.87	1.46	28°	1.0	S	127	0.98	.0183	1.67	(0.0183 * 0.98 * 1.87)/1.67	0.020	20
4	84.0	1.95	1.45	29°	0.5	M	79	5.6	.0184	1.65	2(0.0184 * 5.6 * 1.95)/1.65	0.244	244

Figure 4-63. Permeability Worksheet

2.   Using the Temperature-Viscosity Plot (Figure 4-64), determine Air Viscosity (μ) at the measured temperature and record on the worksheet.

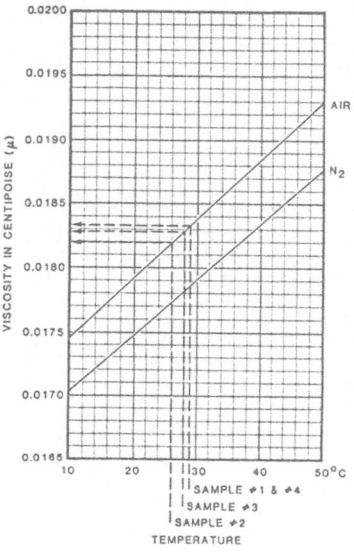

Figure 4-64. Gas Temperature Viscosity Plot

3. Calculate Plug Cross-Sectional Area (A) and record on the worksheet:

$$\text{Cross-Section Area (A), cm}^2 = \pi \left[\frac{\text{Diameter (cm)}}{2}\right]^2 \qquad (4\text{-}18)$$

4. Calculate permeability (K) and record on the worksheet:

$$K \text{ (darcies)} = \frac{M (\mu \, QL)}{A} \qquad (4\text{-}19)$$

where

M = 1, with small flowmeter
2, with medium flowmeter
4, with large flowmeter

$$k \text{ (millidarcies)} = 1000 * K \qquad (4\text{-}20)$$

5. Type a final copy of the Permeability Worksheet (Figure 4-63).

4.35 Saturations

1. Using the Gravity Correction Plot (Figure 4-65), convert the measured gravity of the retorted oil to the estimated original gravity before heating. Record on the worksheet (Figure 4-66).

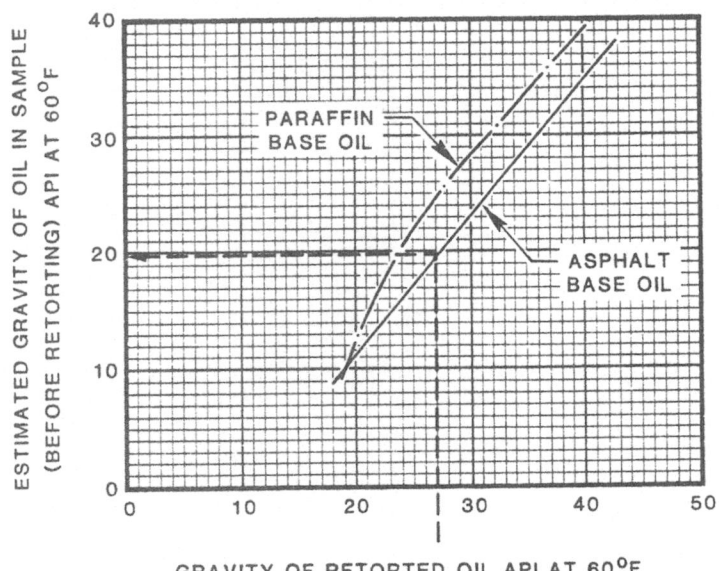

Figure 4-65. Gravity Correction Plot

EXPLORATION LOGGING   SATURATION WORK SHEET   16 00382

Company	XYZ Oil Co	Core	5	Logging Geol	J. Knobla
Well	CAMPBELL #1	Depth	9981-10025'	Date	7-18-81

Pore Volume Sat. Sample (cc) = $\dfrac{\text{Wt. of Sat. Sample w/Fluids (gms)}}{\text{Wt. of Porosity Plug w/Fluids (gms)}}$ × Pore Volume of Porosity Plug (cc)

Water Saturation (%) = $\dfrac{\text{Recovered Free Water (cc)}}{\text{Pore Volume of Sat. Sample (cc)}}$ × 100   Oil Saturation (%) = $\dfrac{\text{Corrected Free Oil (cc)}}{\text{Pore Volume of Sat. Sample (cc)}}$ × 100

Number	Depth	Weight Sat. Sample w/Fluids (gm)	Recovered Free Water (cc)	Recovered Free Oil (cc)	Fm. Oil Gravity API @ 60	Retort Oil Gravity API @ 60	Weight Porosity Sample w/Fluids (gms)	Pore Vol. of Porosity Sample (cc)	Pore Vol. Sat. Sample (cc)	Retort Oil Vol. Corr Factor	Original Gravity Retorted Oil API @ 60	Corrected Free Oil (cc)	Oil Saturation (%)	Water Saturation (%)	Core Water Salinity
							MEASUREMENTS					CALCULATIONS			
1	9981.5	82.51	2.0	0.5	21°	27°	20.74	1.38	5.49	1.2	20°	0.58	10.6%	36.4%	
2	9982.7	83.68	2.2	0.7			21.36	0.99	3.88			0.76	19.6%	56.7%	1100
3	9983.4	94.16	1.7	0.7			19.58	0.62	2.98			0.76	25.5%	57.0%	
4	9984.2	90.77	3.1	0.4	21°	27°	22.43	1.68	6.80	1.2	20°	0.43	6.3%	45.6%	
															1200

Figure 4-66. Saturation Worksheet

2.   Using the Volume Correction Plot (Figure 4-67), convert the Recovered Free Oil Volumes to Corrected Free Oil Volumes and record on worksheet.

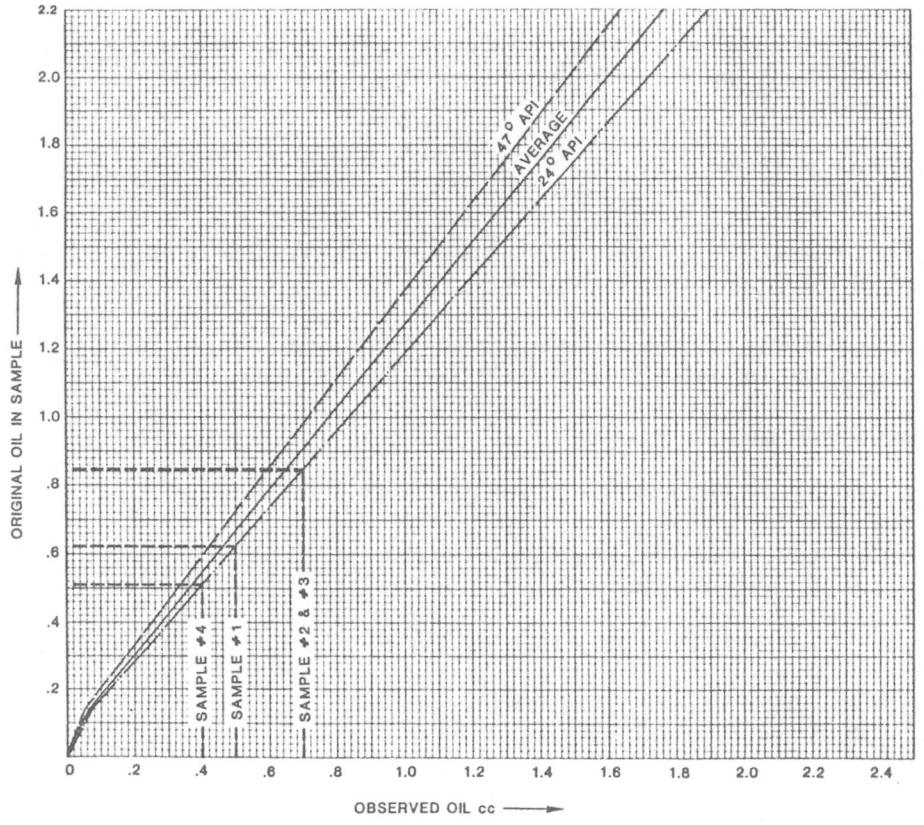

Figure 4-67. Volume Correction Plot

3. Transfer from the Effective Porosity Worksheet the Weight (with fluids) and Pore Volume of Porosity Samples from equivalent depths.

4. Calculate the Pore Volume of the saturation sample and record on the worksheet:

$$\text{Pore Volume (Saturation Sample)} =$$

$$\text{Pore Volume (Pore Sample)} * \frac{\text{Wt (Saturation Sample)}}{\text{Wt (Pore Sample)}}$$

5. Calculate the Oil and Water Saturations and record on the Worksheet:

$$\text{Water Saturation (\%)} = \qquad\qquad (4\text{-}22)$$

$$\frac{\text{Recovered Free Water}}{\text{Pore Volume (Sat. Sample)}} * 100\%$$

$$\text{Oil Saturation (\%)} = \qquad\qquad (4\text{-}23)$$

$$\frac{\text{Corrected Free Oil}}{\text{Pore Volume (Sat. Sample)}} * 100\%$$

6. Type a final copy of the Saturation Worksheet. Add the directly measured formation Oil Gravity and the Core Water Salinity, if available.

4.36 Alternate Methods

4.37 The Oil Correction of recovered oil volume and gravity is required to compensate for the loss of volatiles in distillation.

If for some reason the volume correction graph is not available, or to make a rough check on your results, the following correction factors can be applied.

The average correction factors are:

Low gravity oil (less that °API 20) or asphaltic base oils:

Less than 1.0 cc recovery, multiplying factor	= 1.6
Over 1.0 cc recovery, multiplying factor	= 1.3

High gravity (over °API 20) or paraffin-base oils:

Less than 1.0 cc recovery, multiplying factor	= 1.3
Over 1.0 cc recovery, multiplying factor	= 1.2

If a sufficiently large sample of uncontaminated oil becomes available from a later test, empirical correction factors can be determined for the reservoir by retorting a known volume of oil mixed with clean sand. Comparing the known volume and gravity before the test with the measured volume and gravity of recovered oil, correction factors can be determined.

Charge a retort number with 100 to 120 grams of sand sample completely void of any oil. This sample may be a previously retorted sand or a barren sand from outcrop. The procedure is to add 7 to 8 cc of water to the sand sample and then accurately pipette 2 to 3 cc of the formation oil to the sample. The sample is placed in the still and a retort is run. The oil recovery is carefully measured in the receiving tube.

The volume correction factor is calculated by the ratio:

$$\frac{\text{cc of oil pipetted into sand sample}}{\text{cc of oil recovered after retorting}} \qquad (4\text{-}24)$$

$$= \text{Volume Correction Factor}$$

4.38 The Pore Volume Calculation requires the weight of the porosity sample with its original fluids. This requires that the porosity sample be manually cut from the core. If a core drill is used, the plug will be partially flushed by the cutting fluid, air or water, and its weight cannot be directly compared with that of the saturation sample.

In this circumstance, the pore volume calculation is performed using the ratio of sample dry weights. That is, equation (4-21) is used, but the weights are the extracted and dried weight of the porosity sample and the weight of the saturation sample after it has been removed from the retort.

4.39 Total Porosity may be determined, as previously discussed, if it is required by the client. If this is the case, be sure to discriminate between Total Porosity and Effective Porosity in all written and verbal reports. Casual terminology can cause confusion.

1. Transfer Sample Number, Depth, Dry Weight, and Bulk Volume (V_b) from the Effective Porosity Worksheet to the Total Porosity Worksheet (Figure 4-68).

TOTAL POROSITY WORK SHEET

EXPLORATION LOGGING

Company XYZ Oil Co Core No. 5 Logging Geol. J Knobla

Well Campbell #1 Depth 9981-10025' Date 7-18-81

Number	Depth	Weight Dry (gm)	V_b Bulk Volume (cc)	R_1 Initial Volumeter Reading (cc)	R_2 Final Volumeter Reading (cc)	V_s Sand Grain Volume (cc) R_2-R_1	V_p Pore Volume (cc) V_b-V_s	Ø Porosity V_p/V_b (%) $\frac{V_p}{V_b} \times 100$	Grain Density (gm/cc) Dry Wt. ÷ V_s	Lithologic Description
1	9981.4	19.17	8.78	50	57.4	7.4	1.38	15.7%	2.59	Sandstone, pl gry/grn
2	9982.8	21.30	9.09	50	58.1	8.1	0.99	10.9%	2.63	-red/brn, mod fm, fri,
3	9983.4	18.46	7.92	50	57.0	7.0	0.62	11.6%	2.64	med-f gr, sl slty, arg,
4	9984.3	19.54	8.98	50	57.3	7.3	1.68	18.7%	2.68	subrnd, subelong, frst

Figure 4-68. Total Porosity Worksheet

2. Calculate the Grain Volume (V_g) of the crushed sample from the initial (R1) and final (R2) volumeter readings, and record on the Worksheet:

$$V_g = R2 - R1 \qquad (4-25)$$

3. Calculate the Total Pore Volume (V_p) of the sample from the Bulk Volume (V_b) and the Grain Volume (V_g) and record it on the worksheet:

$$V_p = V_b - V_g \qquad (4-26)$$

4. Calculate the Total Porosity (ϕ) and True Grain Density (ρma) of the sample and record it on the Worksheet.

$$\phi \ (\%) = \frac{V_p}{V_b} * 100\% \qquad (4-27)$$

$$\rho\text{ma} = \frac{\text{Dry Weight}}{V_g} \qquad (4-28)$$

5. Type a final copy of the Total Porosity Worksheet and add brief lithological descriptions abstracted from the Core Report.

6. Using the Total Pore Volume, the Water and Oil Saturations may be recalculated. When reporting these to the client make sure that the different calculation base is understood.

4.40 PRESENTATION OF CORE ANALYSIS DATA

Core analysis data is usually presented in both tabular (the final copy Worksheets) and graphical (the Core Analysis Log) form.

Once all the results and calculations have been collated, a neat presentable report has to be handed to the client.

When using the linear Utility Log sheets (EL P/N 18435 imperial or EL P/N 18436 metric), a graphical representation of the cored interval and analysis results should be drafted as in the example shown in Figure 4-69. Dry transfers should be used for the symbols, heading and labeling, as there is insufficient space in the headings for typewriter lettering. The symbols used should be as shown in the example, and the .35 mm pen should be used for drafting the lithology and the curves. The suggested vertical scale is one major division per foot (two divisions per meter), but a suitable scale should be used to suit the client's needs and to best represent the data.

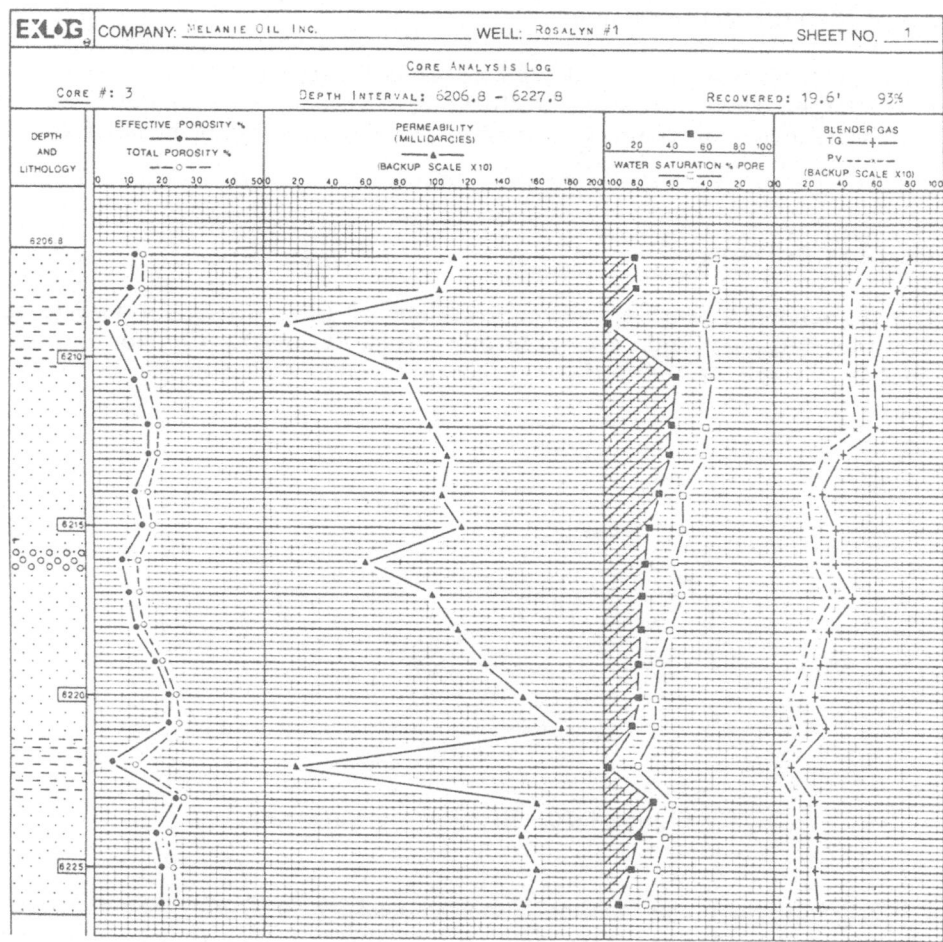

Figure 4-69. Core Analysis Log

The test results should be neatly tabulated by taking fresh worksheets and typing the results as illustrated in Figure 4-61,4-63, 4-66 and 4-68. The same amount of care and attention should be applied in drafting and typing the core analysis reports as is applied in drafting the Master Log.

Accompanying the Core Analysis Log and worksheets should be copies of the Core Report, the relevant sheet(s) of the Formation Evaluation Log, Core and Sample Shipping Inventories.

Procedures for sidewall core samples are essentially the same as for conventional core analysis except for minor modifications, due to the smaller size of the sample. Accuracy is somewhat reduced because of the damage done to the sample by the percussion tool and because of the size of the sample.

Normally, wellsite description of sidewall cores is limited by the requirement to minimize core breakage. However, since performing core analysis on sidewall-cores entails almost total destruction of them, a complete geological evaluation must be performed first.

4.42 CORE DESCRIPTION

Make a 1 mm incision around the circumference of the core about two-thirds of the length from the borehole end, and carefully break it apart. Do not cut completely through, as the passage of the knife may obscure some indistinct structure (see Figure 4-70).

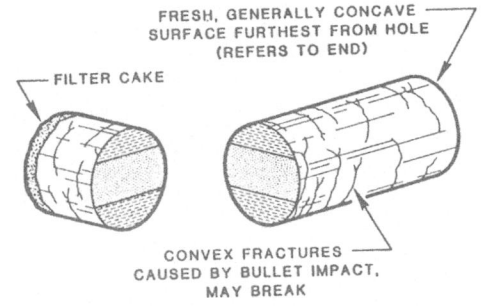

Figure 4-70. Break Core Laterally

Using the shorter piece, make further cuts if necessary to display any sections that are normal and parallel to any structure. The core is now cut such that it may be examined for all relevant possibilities with a minimum amount of breakage (see Figure 4-71).

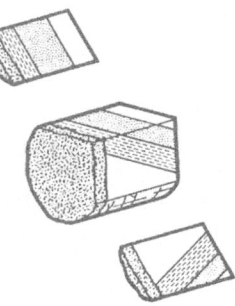

Figure 4-71. Split Shorter Piece Longitudinally

The fresh surfaces allow examination in UV light, without any contamination from lubricants used in preparing the gun. The use of a X10 hand lens is generally faster and more versatile than using the microscope. For detailed characteristics and mineral identification, however, the microscope must be used.

If the core is of a coarse, clastic composition, it will generally be more fragile than one of high argillaceous content. The mechanics of deposition of clastic sediments usually preclude the forming of any sedimentary structure small enough to be recognized in a sidewall core. In more fine-grained sequences, bedding or fine laminae produced by traction carpet mechanisms may be present, but care must be taken if a final commitment has to be made. Ideally, a thorough knowledge of the regional and local geology is necessary to identify a particular sedimentary structure in a clastic rock. A discontinuity may be obvious, but a correct classification can be made only after the consideration of several sedimentary parameters present in the area at the time of deposition, i.e., the position of a sediment and its structure in the Bouma Cycle.

Refer to Formation Evaluation: Geological Procedures (Exlog, 1985) for a discussion of geological techniques and classification systems.

Details of the orientation of the gun in relation to the geological trend is a prime requisite, so that the orientation of any particular structure may be ascertained to facilitate correct interpretation. This is most important in the interpretation of structures in any rock.

Don't confuse epigenetic structures with primary sedimentary structures. Pressure solution cleavage, grain-welding anisotropies, mica and clay segregations, concretion boundaries, veins, bioturbation, ghost structures from dissolved fossil remains, and fractures are some major examples. With the limited equipment provided in most logging units, a correct interpretation of one particular structure would be almost impossible; however, logical elimination may narrow the possibilities to two or three structures from which an intelligent interpretation can be made. If this is not feasible, then make a short list of possibilities on the report.

Identifying the mineral composition and bulk percentages is important, particularly when correlating porosities and permeabilities from the wireline logs. Heavy mineral sands, tuffs and generally clastic rocks that have a composition different from the classic quartz-clay rock will produce a response that may be incorrectly interpreted if the drilling engineer is not aware of unusual geological parameters. The percent concentration of heavy minerals in a sediment is rarely correctly estimated during normal logging procedures. This is primarily a function of their rarity in that they tend to be overlooked in the sample tray, and also because their high density causes higher sink velocities in the annulus. This increases the possibility of their 'hanging up,' particularly in a long, large-diameter riser used in some offshore wells. However, in thick sands or tuffs that have high concentrations of heavy mafic minerals, their concentration may be estimated during mud logging -- but it will generally be a very pessimistic representation of the true amount.

If the core is wholly or predominantly argillaceous, it was taken mainly for source rock analysis and biostratigraphic control. To obtain most information from a clay core it should be carefully prepared to display any structures to their best advantage. When examining and describing clay cores, do not take the attitude that, visually, one and all are very similar -- and then, after the first two or three have been described, carelessly report the rest as being precisely the same unless there is something very obvious to catch the eye. Unfortunately, analysis of mineral composition of a clay core is not possible in an average logging unit, and visual appraisal of grain-to-grain characteristics is also precluded by their size. This, however, is no real disadvantage: as a legacy of the submicroscopic grain size, structures are consequently on a much smaller scale in comparison to arenaceous sediments. The majority of predominantly clay sediments were deposited on a horizontal sediment-water interface, such that the mechanics of deposition in relation to the average clay particle shape caused all particles to be deposited with their long axes parallel to the sediment surface (horizontal). If the clay is bentonitic in character it will not possess this finely laminar structure as a primary sedimentary feature, but may later acquire fissility through tectonic processes. Chloritic, illitic, and kaolinitic clays will preserve their original sedimentary structures (unless they are physically disturbed) because of their nonhydrating capacities. Montmorillonite clays usually have no original sedimentary structures due to the effects of dehydration which, when initiated, rapidly redistributes ions -- causing virtual mass-movement.

The extremely slow rates of deposition of clays allow a very small environmental change to be recorded on a very small scale, in comparison to clastic sediments. On the scale of the sidewall core, hundreds of years of deposition may be represented in the thickness of the diameter of the core. Within that time-scale, subtle changes in climate, current, source area, water temperature, water depth, and zoological activity may be recorded as a change in character of the clay on the seafloor at that time. Conditions may fluctuate, cycle, or continually change such that changes in clay composition and structure may be either gradual or very rapid with depth. Diligent examination of the clay cores, carrying out the same procedures for each one and carefully noting the results, may reveal geological information relating to the paleo-environment or source that would have been overlooked if they have been given only cursory attention.

The orientation of planar structures (bedding laminae, veins and cleavage) is particularly useful in that the local geological trend may be determined, if the attitude of the bullet was known, and in conjunction with the use of an isopach map. In near vertical wells the bullets enter the formation almost horizontally. Thus any dip noted in any laminar structure in a clay sediment may be taken to represent the local dip of the formation, unless the particular structure is known not to be related to original sedimentary processes (veins).

4.43 CORE ANALYSIS

4.44 Sample Preparation

The sidewall sample needs only to have the mud cake scraped off with a knife or brush to be ready for analysis. Do not crush the sample for saturation determination, because the entire analysis is run on one sidewall sample plug. Sidewall cores will normally require wax-mounting for the permeability test. The exception is when the plugs are sufficiently firm to be shaped with a knife to fit the R-20 permeameter sleeve.

4.45 Porosity and Permeability

1. Describe the core and scrape mud cake from it. Check for fluorescence under the ultraviolet light.

2. With a knife, shape the core for porosity and permeability tests. The same test sample is used for both tests, with the porosity test run first.

3. Weigh the test sample with fluids, to the nearest 0.01 gram.

4. If there is no fluorescence, dry sample in the shelf oven for approximately one hour. If there is fluorescence, then it will have to be extracted.

5. Weigh sample without fluids, to the nearest 0.01 gram.

6. Determine bulk volume with porometer.

7. Determine compression volume with porometer and convert to grain volume.

8. Calculate effective porosity.

9. Shape test sample to cylindrical dimensions to fit R-20 rubber sleeve, or, if unconsolidated, for mounting in wax.

10. Measure length and diameter of sample with Vernier caliper to nearest 0.01 cm.

11. Mount sample plug in appropriate rubber sleeve, depending on whether or not sample is mounted in sealing wax.

12. Run permeability test. Be especially cautious of cracks in the test sample which will give incorrect, high-permeability values.

13. Run core chip salinity.

4.46 Saturation

Sidewall cores are taken from the zone of mud invasion and flushing. Saturations in this zone are of little value. However, if they are required, the saturation test is run first on the entire plug, so do not crush the plug into chips. After the saturation test is run, the sample is shaped for the porosity and permeability tests.

Calculation of saturations is similar to the method described in Paragraphs 4.35 and 4.38, using the dry weight ratio.

4.47   CARE OF THE CORE ANALYSIS KIT

4.48   CLEAN UP AND STORAGE

1. Clean mercury off the outside of the porometer and tray; collect used mercury in a beaker topped up with water and detergent. Clean the sample chamber with a cotton swab. Clean the surface of the mercury in the pump. Clean the porometer lid of mercury and dirt; remove the O-ring, carefully clean the ring and seat, and replace; change the O-ring if it appears worn or damaged. Clean the inner seat of O-ring in the chamber. Tighten the porometer lid, run the mercury in about half-

way (20 cc), close the needle valve, close the valve to the low pressure gauge, and lock the handle by tying, taping or using a nail.

WARNING

Mercury is poisonous. Don't leave it lying around: keep it away from heat; keep it off your clothes. Always wear rubber-soled shoes to perform core analysis. Clean up all spilled mercury immediately! Use rubber bulb or eye dropper for this.

2.   Clean up all glassware, and store it in the proper drawer. Glassware and everything else used for core analysis should be spotlessly clean. Clean all retort bombs, inside and out, and store away neatly. Replace asbestos gaskets in the lids when necessary (if worn or damaged). Clean out stills with a brush dipped in acetone or chlorothene (wait until they have completely cooled off!).

3.   Clean wax and sand from the board used for mounting samples. Once the permeability tests have been run, mounted plugs can be removed from the metal sleeves by heating the outside of the sleeve with a torch or soldering iron; the plug will slip out of the sleeve. Deliver plugs to the client. Clean the inside of the rings, and remove all traces of wax with chlorothene or acetone.

4.   Put everything away neatly in its place so that you or anyone else can find it again without difficulty.

Remember: Be clean and careful! Strive for accuracy first, speed second.

4.49   TEAR-OUT AND SHIPPING

The following are the normal packing and shipping instructions for the core analysis equipment discussed in this manual.

Porometer: Bolt securely in its wooden crate. Raise pressure in the lower pressure gauge to 30 psi and close the valve to this gauge. Next, tape the crank handle to prevent its rotation. Nail down the top of the wooden crate.

Permeameter: Remove from wall and bolt it to the shipping bracket. Replace the four wall screws and washers in the wall of the unit. Remove the yellow centigrade thermometer from the permeameter and package it with the spare thermometer. Carefully pack and secure the permeameter in its shipping crate.

Saturation Stills: Remove stills from the wall and replace the mounting screws back in the wall. Be sure to remove the retorts during shipping. Package the stills in their cardboard shipping case. Double-check to see that electric cords have not been left in the unit.

Refer to the core analysis parts inventory list when packaging the remaining parts for shipment, checking off each item on the list as it is packed. Any missing or damaged items should be listed on the Final Request for Supplies and Operating Record. This form should be sent to the local office, explaining how, when, and where the equipment was shipped. Be thorough and <u>do not</u> leave any parts or manuals in the unit. At this time, the core analysis equipment set should be completely ready for the next job. Any repairs or replacements required should be according to your report. It should not be necessary for the next crew or a service engineer to clean up or make wellsite repairs prior to next using the kit.

If the Ruska core analysis equipment is damaged beyond repair or is malfunctioning, it will have to be replaced in the field. As the porometer, permeameter, and saturation stills are self-contained units, their replacement is basically a simple exchange. <u>It is extremely important that the Ruska manuals be exchanged along with the equipment.</u> The reason for this is that the factory calibration of the porometer and the permeameter are applicable <u>only to one specific instrument.</u> The serial numbers and the instrument should <u>always</u> be identical.

INDEX